LOST KNOWLEDGE of the ANCIENTS

"These accounts of 'lost' or anomalous knowledge are building blocks of an alternative universe, and, taken altogether, they reveal something of the deep structure of this alternative universe. They also show that this universe made up of anomalies may be more real than the everyday, commonsensical one!"

MARK BOOTH, AUTHOR OF
THE SECRET HISTORY OF THE WORLD

LOST KNOWLEDGE
of the ANCIENTS

A GRAHAM HANCOCK READER

EDITED BY GLENN KREISBERG

Bear & Company
Rochester, Vermont • Toronto, Canada

Bear & Company
One Park Street
Rochester, Vermont 05767
www.BearandCompanyBooks.com

Text paper is SFI certified

Bear & Company is a division of Inner Traditions International

Library of Congress Cataloging-in-Publication Data
Lost knowledge of the ancients : a Graham Hancock reader / edited by Glenn
Kreisberg.
   p. cm.
  Summary: "Thinkers at the forefront in alternative theories on history, the
origins of civilization, technology, and consciousness"—Provided by publisher.
  Includes bibliographical references and index.
  ISBN 978-1-59143-117-6 (pbk.)
  1. Science and civilization. I. Hancock, Graham. II. Kreisberg, Glenn.
  CB478.L67 2010
  303.48'3—dc22

           2010014856

Printed and bound in the United States by Lake Book Manufacturing
The text paper is 100% SFI certified. The Sustainable Forestry Initiative program
promotes sustainable forest management.

10 9 8 7 6 5 4 3

Text design by Jon Desautels and layout by Priscilla Baker
This book was typeset in Garamond Premier Pro with Stone Serif and
  Copperplate used as display typefaces

Images 10.1 and 10.4 are provided courtesy of www.SacredSites.com.
Images 14.1 through 14.10 are provided courtesy of Santha Faiia at
  www.GrahamHancock.com.
Additional images are reprinted from www.GrahamHancock.com unless otherwise
  noted.

To send correspondence to the author of this book, mail a first-class letter to the
author c/o Inner Traditions • Bear & Company, One Park Street, Rochester, VT
05767, and we will forward the communication.

I would like to dedicate this work to my wife,
Stephanie, son, Jason, and daughter, Sophie,
and thank them for offering their support,
love, and encouragement in our search,
as a family, for new truths.

Contents

Acknowledgments

I came to the idea for this compilation with the cooperation of author/researcher Graham Hancock and the contributors to his online forum at www.GrahamHancock.com. This volume can be considered a select offering from the many dozens of excellent articles and essays on the site that speak to the search for lost knowledge and the knowledge of the great astronomical cycles of our existence.

I would like to thank and acknowledge the kind friendship and support of Graham Hancock, John Anthony West, Robert Schoch, and Walter Cruttenden of the Binary Research Institute (BRI), and all of the contributing authors who cooperated in putting this work together. Together, we offer these essays with a commitment to the principles of embracing the pursuit of knowledge and truth through rigorous, open-minded research, discovery, and critical analysis.

On the Search for Lost Knowledge

GLENN KREISBERG

A revolution is underway in our world, fostered by decades of research and well documented evidence from illustrious scholars such as Robert Bauval, Graham Hancock, John Anthony West, and Robert Schoch (sometimes affectionately referred to as the "four horsemen"). This collection of essays adds to those voices and, when taken as a whole, provides astounding, yet compelling proof of an approaching shift in our perspective regarding human origins and the roots of civilization. Long-standing paradigms are beginning to crumble and realign to this new order of thinking.

This "thought revolution" represents a direct challenge to an orthodox, mainstream academic establishment that continues to try to debunk sound evidence and theories that have not only stood the test of time but that have already attained accepted status in the court of public opinion. Of course, orthodoxy resists and fights change by its very nature. Acceptance by this establishment would require a shift in perception regarding the knowledge that mankind possessed in remote epochs.

Why lost knowledge? Because much of the research, evidence, and theories presented in this collection are not currently included in mainstream, traditional academic curriculums. Nor are they seriously pursued at universities or published in scholastic textbooks and journals, although that is slowly changing. The battle to have them included and accepted has stirred controversy worldwide for decades, in some cases prompting attacks by academia against those involved in this fight.

Lost Knowledge of the Ancients offers the reader a selection of well researched and thought provoking articles and essays covering alternative science and history. We have tried to stay within the broad theme of lost knowledge and cycles of time, while exploring nontraditional theories and perspectives that challenge long-standing assumptions. The diverse offerings of these prolific authors and researchers present a unique snapshot of some of the latest findings in cutting-edge scientific and historical research, while exploring new theories and fresh challenges to existing ones.

By recognizing that many modern discoveries are actually rediscoveries of lost knowledge from past civilizations, the essays in this collection aim to contextualize, within the bigger picture of human knowledge, the achievements and events giving rise to a fundamental understanding of where our present time fits into the past and the future; into the cycle of human existence.

These essays also connect the dots, so to speak, for readers who will discover that the whole picture adds up to more than the sum of its parts. Through these writings, readers will realize that many of these authors, scientists, and researchers suggest that there is a cycle of time and consciousness that profoundly affects the way we look at history. Each article in this collection articulates something that builds on the overall theme.

For example, Dr. Schoch looks at possible higher states of consciousness (telepathy, clairvoyance, and so on) in a past age, not just because he is interested in the paranormal, but also because it adds to and builds upon his earlier work. He finds that these phenomena have real scientific support that buttresses the ancient stories of a pre-Babel era when mankind communicated at a higher level, allowing such things as building great structures long before there was any form of written language (heretofore considered a prerequisite to "civilization"). Edward F. Malkowski, in his essay, looks deeply at the wisdom and cosmology of ancient Egypt and finds a parallel to modern string theory and quantum physics, which wouldn't be considered possible under the linear paradigm of history.

So, as the paradigms of science and history shift (virtually on a daily basis), this committed breed of scientific researcher endeavors to break new ground and report on remarkable discoveries and astonishing new theories and discoveries in the fields of science, history, philosophy, and religion. Committed to standards of scientific methodology, this collection brings to the inquisitive reader the most relevant, insightful, and startling scientific and historical information published in our world today.

These authors rarely write outside their fields. In fact, some of them may not even fully realize how their work fits into the bigger picture—but by fitting all the pieces together, we now have enough knowledge and evidence to tentatively confirm that what the ancients themselves told us is real: Time really does move in a cycle, and the myths related to this cycle reveal both a long-lost Golden Age, and a new one that approaches!

We begin with an article by Dr. Robert Schoch, a geologist and full-time faculty member at Boston University since 1984, who is best known for his controversial redating of the Sphinx from 2500 BCE to at least 5000 BCE (and perhaps even much earlier), based on sound geologic evidence. Entering another field, Dr. Schoch writes on examining evidence and research surrounding parapsychology and "supernatural" phenomena, shedding new light on the process of evaluating such research and evidence.

Author/researcher Edward F. Malkowski writes about the connection between the ancient Egyptian mystery school and today's new science philosophy, showing that "new" revelations actually shadow ancient insights.

Explaining and expanding his Orion Correlation Theory, researcher/author Robert Bauval demonstrates how a "Cosmic Egypt," ghosted in the geography of the Nile Valley, reflects the cyclical changes in the sky landscape caused by the precession of the equinoxes. Bauval shows how this influence over the evolution of temple building, for over three thousand years and over a thousand miles, was administered by astronomer/priests of the highest order.

Author Mark Booth uncovers the hidden truths behind the esoteric knowledge held and passed down by secret societies throughout history, and in doing so opens our eyes to a new way of looking at the history of the world.

Famed rebel Egyptologist John Anthony West takes on the "Church of Progress" and calls to task the "mainstream" archaeological community in his open letter to the editors of *Archaeology* magazine.

Secret societies also form the basis for Richard C. Hoagland and Mike Bara's fascinating research into the possibility of an ancient race once inhabiting Mars. Hoagland claims that NASA is covering up evidence found on both the moon and Mars that proves a highly advanced, yet ancient, civilization once existed and may hold the key to the lost history of the human race.

Walter Cruttenden, founder of the Binary Research Institute (an archeoastronomy think tank), explores what Plato described as the "Great Year," a 25,000+ year cycle based on the precession of the equinoxes, which may account for the rise and fall of civilization as well as human consciousness. Cruttenden shows how the "myth" associated with the precession cycle pervaded all ancient cultures.

Continuing on the theme of a grand cycle of time, researcher Scott Creighton offers a convincing new theory accounting for the ground plan at Giza, Egypt, linking it to a warning of cosmic destruction encoded into a precessional calendar wheel clearly laid out among the temples, monuments, and pyramids of the Giza Plateau.

Researcher Andrew Collins postulates that a cosmic burst of energy from a relatively nearby star during Paleolithic times may have boosted human consciousness and kick-started human religion and creativity by affecting changes in our evolution.

Speculating that the ancients may have known much more than they are given credit for, I examine the evidence that may prove that our ancestors had knowledge and perhaps use of the electromagnetic spectrum in ancient times.

Badrinaryan Badrinaryan, chief geologist for the National Institute for Ocean Technology in India, takes us to the floor of the Gulf of Khambhat, off India's northwest coast. Evidence from side-scan sonar images and artifacts brought up from the murky depths helps build the case for a new cradle of ancient civilization and offers us a glimpse of what the oceans are likely to give up to archaeological explorers during the next century.

Author Gary A. David reveals a remarkable correlation between the stars of the constellation Orion and the Anasazi ground constructions in the southwest United States, not unlike the correlation found in Egypt. These careful astronomers meticulously laid out both winter and summer solstice sunset and sunrise alignments as part of a "ley" energy system covering their territory from southern Colorado through Arizona, to the mouth of the Colorado River.

Italian naval engineer Flavio Barbiero makes the case for the possibility of instantaneous shift of the poles; a cataclysmic event that may have occurred in relatively recent times, destroying a once flourishing civilization whose ruins, providing evidence of the shift, now lie buried under the ice sheets of Antarctica.

And last, Graham Hancock offers us startling new evidence about those who first peopled the Americas and where and when they came from.

In all, it's a fascinating collection of articles and essays meant to widen our perspective, expand our horizons of knowledge, and speak to the mysteries that drive the human heart and mind to explore and discover new truths about the human condition we all share.

At present, it's being reported around the world that former NASA astronaut Dr. Edgar Mitchell, who flew the *Apollo 14* mission and was the sixth human to walk on the moon, announced that the United States government has been covering up a sixty-year-old incident involving human contact with intelligent extraterrestrial aliens. Dr. Mitchell holds doctorates of science in astronautics and aeronautics and worked at NASA for decades. He claims his report is based on inside information and years of firsthand evidence. This can't be easily disputed, nor can it be dismissed out of hand—yet it fails to gain any real traction as a credible news story in today's world.

NASA, the "official establishment," issued a response disagreeing with Dr. Mitchell's "opinion." NASA did not acknowledge that Dr. Mitchell was not expressing an opinion, he was relating a fact. Along with many of the contributors to this book, he may be proven right over time, simply by the facts they've stated. Fifty, one hundred, five hundred, or a thousand years from now, when historians are writing the history of our present time, they will no doubt acknowledge the few visionary researchers and writers who got it right and worked to bring forbidden knowledge and new insights to the attention of the world.

Thoughts on Parapsychology and Paranormal Phenomena

HAS CONVENTIONAL SCIENCE MISSED SOMETHING PROFOUND?

ROBERT M. SCHOCH, PH.D.

Many of you reading this piece may know me from my work on the Great Sphinx, the Great Pyramid, and other ancient monuments, as discussed in various articles, documentary videos, and the books that I have written. As such, I am commonly asked how my studies in parapsychology relate, if at all, to my studies of ancient monuments. Even though I am a geologist, and my initial concern is in dating various ancient structures, I can't help but wonder why they were built, especially given the enormous efforts that must have gone into their construction.

The "why" behind the monuments, more often than not, apparently included motivating factors such as religious beliefs and practices, initiation rites, and rituals, which in many cases seemed to have an ostensible paranormal aspect, whether it was clairvoyance, divination, or manifestations of higher levels of consciousness. The temples and tombs of ancient Egypt, Mexico, and Peru seem to cry out "paranormal." So, was it all a mixture of ancient myth, superstition, and downright fraud on the part of many a seer, seeress, priest, and priestess, or could there have been something to it? Were the ancient structures used, at least in part, to alter consciousness, and possibly enhance paranormal phenomena?

In all honesty, I have always been highly skeptical of any alleged paranormal phenomena. However, my concept of skepticism is not the same as dismissal, and in my studies of ancient, indigenous, and traditional cultures, alleged paranormal phenomena kept making an appearance. When Logan Yonavjak, a former Boston University student of mine, encouraged me to delve deeply into the serious parapsychological literature, I found the topic both fascinating and enlightening. The immediately tangible result of our research was the book we wrote together entitled *The Parapsychology Revolution,* in which we included selections from fourteen seminal papers, dating from 1886 through 2007, by major figures in the field, as well as a hundred pages of our own commentary. For me, however, the real result of my immersion into parapsychology was a new appreciation for human potentiality and the connections we share with all of life and ultimately, perhaps, with the cosmos.

Are Paranormal Phenomena Real?

People often ask me if I "believe" in the paranormal. Let me just say that after looking at the hard evidence, and sifting out the fraud and bunk, I have come to conclude that there definitely is something to such phenomena as telepathy and psychokinesis.

Most people who have seriously studied the subject conclude that telepathy (mind to mind interaction) is the best-supported class of paranormal phenomena. There is strong laboratory evidence for telepathy, such as classic card-calling experiments, as well as many more sophisticated tests of telepathy, clairvoyance, and remote viewing. There is also a large and compelling body of evidence from spontaneous cases supporting the reality of telepathy. For instance, crisis apparitions, veridical hallucinations, or "ghosts" are well known, as documented in the classic two-volume scientific monograph of rigorously authenticated events produced by the Society for Psychical Research titled *Phantasms of the Living*.

The evidence for psychokinesis is also strong, including micro-psychokinesis studies using random event generators and similar devices. Examples of this evidence include the work done by the Princeton Engineering Anomalies Research (PEAR) labs over more than a quarter of a century, and the carefully studied incidents of macro-psychokinesis associated with genuine spontaneous poltergeist cases. Another line of compelling evidence for the reality of paranormal phenomena is the study of presentiments or "pre-sponses," essentially a form of short-term precognition as measured by physiological parameters (heart rate, electrodermal activity, and so forth). Numerous replicated experiments have demonstrated the physiological responses of individuals to, for instance, disturbing photographs a second or two before they are actually viewed by the person. According to conventional science, this should not be possible.

As a natural scientist, I expect genuine phenomena (be they psychic and paranormal, or more conventional phenomena) to exhibit patterns and share elements in common, and this is just what has been found in spontaneous cases of the paranormal. Even when viewed cross-culturally, such commonalities persist.

Perhaps even more compelling for me is the work of various

modern researchers that has demonstrated a weak but persistent correlation between low levels of geomagnetic activity on planet Earth and cases of apparent spontaneous telepathy (based on records going back to the latter half of the nineteenth century). This, in my opinion, is a very strong argument supporting the contention that there is something genuine to the concept of telepathy. It suggests that spontaneous telepathic phenomena are real and natural and, as might be expected of natural phenomena, that their manifestation is influenced by other natural parameters.

Alternatively, are we to hypothesize that hundreds of hoaxers over nearly a century and a half have conspired to fake telepathic incidents in identical correlation with geomagnetic activity? This latter hypothesis strikes me as rather far-fetched, if not downright ludicrous. It has also been found that incidents of the paranormal correlate with local sidereal time (which relates to the position of the horizon at any particular point on Earth relative to the center of our galaxy).

Note that a correlation between geomagnetic activity and spontaneous telepathy does not necessarily imply that the "telepathic signal" is magnetic or electrical in nature. The human brain is influenced by magnetic and electric fields, and whatever may be the carrier of the telepathic signal, the transmission, reception, and manifestation of the message by the brain could be hampered or enhanced by differences in the magnetic and electric fields that the brain is subjected to.

For many people, a phenomenon is not "real" unless it can be duplicated in a laboratory setting under controlled conditions. Being a natural scientist and field geologist, I have never agreed with this contention. After all, can we create a genuine volcanic eruption in the laboratory or even on command in the field? Until about two centuries ago, the scientific community routinely rejected the concept of rocks falling from the sky (meteorites). Still, attempting to induce, capture, observe, and experiment with apparent telepathy under controlled conditions is a worthy endeavor. Unfortunately, however, it is fraught with problems to this day, and though numerous experiments have tested positive for apparent telepathy, others have had negative results, and replication is a persistent problem. The bottom line is that we really do not know exactly what parameters or variables make for good telepathic transfer (or the elicitation of other types of paranormal phenomena), much less how to control for them.

There are major issues that remain unresolved concerning paranormal and psychic phenomena. We don't fully understand what conditions are best to elicit paranormal phenomena, and thus these phenomena are not easily replicated on command (such as in a laboratory setting). There is often a very low signal to noise ratio when it comes to psychic phenomena; there is no single physical theory to account for paranormal phenomena; and there is the issue of fraud and charlatans. Fraud is a very real and persistent problem in the field of psychical research, and one of the reasons to undertake large statistical studies of average people (as opposed to so-called psychic superstars) and search for the regularities and patterns one would expect among any genuine natural phenomena.

Also, paranormal studies have extended to animals (and in some cases, even plants). One of the strengths of nonhuman studies is that it is highly improbable that animals will cheat and lie. It can also be noted that many "powerful mediums" who appear to have genuine paranormal abilities also apparently have low moral values and will cheat and commit fraud (perhaps unconsciously) at times, especially when their genuine paranormal powers fail. This is a pattern that has been noted over and over among parapsychologists working with human subjects.

The Human and Cosmic Psyche

Speculations on Precessional Ages,
the Yuga Cycle, and the Mental Abilities
of Ancient Peoples

Many cultures and societies have traditions that hold that history is cyclical, which is expressed in numerous ways. For example, the classical notion of Gold, Silver, Bronze, and Iron Ages; the sequence of the twelve zodiacal precessional ages that make up the Great Year; or the yuga cycle of Hindu philosophy. What, I have to ask myself, if there is some truth to such cycles? Has human consciousness, perhaps expressed in terms of different mental (and paranormal) abilities and priorities, varied with such cycles? If so, can we discern evidence of such changes in the historical and prehistoric record? Initially I was highly skeptical that there could be any reality to "world ages," but as I study the data, I'm not so sure.

The evidence of the ancient monuments, combined with the mythology* clearly indicates that a knowledge of the precessional cycle[†] was well known throughout the ancient world as far back as five thousand years ago, and maybe much earlier. Accounts in the Judeo-Christian Bible, motifs of Egyptian art and architecture, and themes found among Greco-Roman myths (to name just a few examples) reflect precessional changes.

We are now in the end of the Age of Pisces, the age that began some two thousand years ago, not merely coincidentally with the birth of a new religion (Christianity) that adopted the fish (Pisces) as one of its earliest symbols. Soon we will enter the much-ballyhooed Age of Aquarius. Working back in time prior

---

*Myths are not simply fairy tales or false stories, and in ancient times often encoded both pragmatic information and profound knowledge.

[†]The precessional cycle is comprised of the Age of Aries, the Age of Pisces, the Age of Aquarius, and so on, based in the West on the location of the sun among the zodiacal constellations on the vernal equinox.

to the Age of Pisces, we have the Ages of Aries, Taurus, Gemini, Cancer, Leo, Virgo, Libra, Scorpio, Sagittarius, and Capricorn, and then an earlier Age of Aquarius (and the entire cycle repeats again, and again, and again).

Hindu tradition has the yuga cycle, consisting of Satya Yuga (also known as the Krita Yuga; roughly a Golden Age), Treta Yuga (Silver Age), Dwapara (Dvapara) Yuga (Bronze Age), and Kali Yuga (Iron Age). (Note that these ages in the Hindu system vary in length among themselves.) In the late nineteenth century, Swami Sri Yukteswar Giri suggested that one entire yuga cycle takes approximately 24,000 years (descending from the Satya Yuga down to the Kali Yuga, and thence ascending through the yugas to another Satya Yuga, a pattern which repeats over and over) and is tied to the precessional cycle, which, in turn, is due to our sun revolving around another star. According to Sri Yukteswar, Earth left the last Kali Yuga circa 1699 CE, and we are thus currently in the first centuries of a Dwapara Yuga. Furthermore, he also asserted that Earth is subjected to varying magnetic fields and other influences and energies as it travels through space, affecting the mental attributes of those who inhabit the planet.

Approximately 15,000 years ago, Earth was at the beginning of the precessional Age of Virgo (using the Western classical system, and depending on the exact point used to separate the Age of Virgo from the preceding Age of Libra). Using Sri Yukteswar's dating of the yugas, 15,000 years ago Earth was in the Satya Yuga ("Golden Age"). What was happening 15,000 years ago on Earth? It was the end of the last ice age, but that did not mean the entire planet was covered in ice—many areas were still temperate to tropical. Our ancestors, modern humans (*Homo sapiens*) inhabited the planet, but we were not alone in terms of human and humanlike species.

The diminutive species *Homo floresiensis* lived in what is now modern Indonesia. An adult *H. floresiensis* stood about three and a half feet tall and had a brain only the size of a modern chimpanzee brain (not much larger than one-fourth the size of the modern human brain on average). Yet, despite their small brains overall, there is evidence that the portion of the brain associated with self-awareness is approximately the same size in *H. floresiensis* as in *H. sapiens*. Self-awareness, it can be argued, is an attribute that one would expect both in the Golden Age of the Satya Yuga and the precessional Age of Virgo since Virgo, ruled by Mercury, is associated with mental activity.

Fifteen thousand years ago in southern Africa lived another species, technically known as *Homo capensis,* so-named in 1918 on the basis of a partial skull (note that there is disagreement as to whether or not these forms represent a species distinct from us). More commonly they are referred to as the Boskop people or Boskops. Though their bodies were about the same size as those of modern humans, Boskops had huge heads with brains 25 to 35 percent larger than those of modern humans. In their book *Big Brain: The Origins and Future of Human Intelligence,* neuroscientists Gary Lynch and Richard Granger argue that Boskops were in fact more intelligent than modern humans. Increased brain activity and enhanced levels of consciousness (perhaps expressed in manifestations that we would deem as paranormal) may have characterized the Boskops; this would seem to fit well with both the Satya Yuga and the Age of Virgo. Is this just coincidence? Or is this correlation significant?

Science and Paranormal Phenomena

A popular approach to possible paranormal phenomena is to simply dismiss such things as impossible—impossible either in an absolute sense, or as being of such a "low probability" as to be unworthy of consideration. For example, Sean Carroll, senior research associate in physics at the California Institute of Technology, posted a diatribe against parapsychologists and the very idea of studying possible paranormal phenomena on his blog at *Discover* magazine's "Cosmic Variance" site. The core of his argument is as follows: "The main point here is that, while there are certainly many things that modern science does not understand, there are also many things that it *does* understand, and those things simply do not allow for telekinesis, telepathy, etc. Which is not to say that we can *prove* those things aren't real. We can't, but that is a completely worthless statement, as science never proves anything; that's simply not how science works. Rather, it accumulates empirical evidence for or against various hypotheses." [Italics in the original.]

I'm classically trained in the sciences, and I understand where Carroll is coming from philosophically when he states that "science never proves anything" (and I agree that, epistemologically, at a deep level, proof is not the domain of science). However, Carroll contradicts himself when he states: "The crucial concept here is that, in the modern framework of fundamental physics, not only do we know certain things, but we have a very precise understanding of the *limits of our reliable knowledge.* We understand, in other words, that while surprises will undoubtedly arise (as scientists, that's what we all hope for), there are certain classes of experiments that are guaranteed not to give exciting results—essentially because the same or equivalent experiments have already been performed." [Italics in the original.]

Here Carroll is clearly, and for all practical purposes, arguing that certain things have been "proven" when he asserts that "we know certain things." This strikes me as a modern version of the famous (and famously wrong) 1894 pronouncement made by Albert A. Michelson (later to become a Nobel laureate in physics), quoted in an article in *Natural History* by Neil deGrasse Tyson titled "The Beginning of Science": "The more important fundamental laws and facts of physical science have all been discovered, and these are now so firmly established that the possibility of their ever being supplanted in consequence of new discoveries is exceedingly remote. . . . Future discoveries must be looked for in the sixth place of decimals."

Michelson made this statement before the elucidation of X-rays and the structure of the atom, and before the discovery of radioactivity and the development of quantum physics and relativity theory. Despite his scientific brilliance, Michelson did not prove himself a very good diviner of the future.

Returning to the quotation from Sean Carroll, he states: "The main point here is that, while there are certainly many things that modern science does not understand, there are also many things that it *does* understand, and those things simply do not allow for telekinesis, telepathy, etc."

If in place of "telekinesis, telepathy," we substitute "continental drift," I could easily imagine this statement being made in the early twentieth century. The very concept of moving continents was lambasted from some quarters, despite the strong evidence in support of the theory, because it was deemed "impossible" and utterly "inconceivable" based on the science of the time that continents could move. There was no known mechanism, until, that is, the development of plate tectonic theory. (Yes, as a geologist, I am aware of the various criticisms and possible shortcomings of tectonic theory, but that is not the point or issue here.)

In analyzing further the selected quotations from Sean Carroll, we see that they actually expose the weaknesses of like-minded individuals who insist that science in its modern Western guise should have the last word or, to put it another way, be the final arbiter of truth (even if only a provisional truth). Again, the Carroll quotation is: "The main point here is that, while there are certainly many things that modern science does not understand, there are also many things that it *does* understand, and those things simply do not allow for phenomena such as telekinesis, telepathy, etc. Which is not to say that we can *prove* those things aren't real. We can't, but that is a completely worthless statement, as science never proves anything; that's simply not how science works. Rather, it accumulates empirical evidence for or against various hypotheses."

To say that something is "worthless" is a value judgment, and in an ultimate sense, one can argue, value judgments are outside of the realm of science and dependent on an emotional investment and context, among other factors. I would argue that the fact that "science never proves anything" is quite valuable as it precludes the "sleight-of-hand" dismissal of the very possibility of any supposed paranormal phenomena ever being genuine.

As far as empirical evidence is concerned, for many people this is the heart of the issue when it comes to alleged paranormal phenomena. Exactly what is evidence? Are anecdotal case studies evidence? Is it appropriate to pick and choose what we consider evidence and dismiss, as "nonevidence," case histories and studies that do not suit our biases? Is it legitimate for certain critics and debunkers to reject as evidence experiments carried out by particular laboratories, which give positive results for telepathy (with good controls, and taking all possible precautions against possible fraud)? Yet these same critics and debunkers will hail as "solid evidence" against the telepathic hypothesis similar experiments in a comparable lab, which give negative results (whatever "comparable" is in such cases, given how poorly we understand the aspects that influence paranormal phenomena).

In fact, what is and is not "empirical evidence" is not a black and white matter, either in parapsychology or in many other scientific disciplines. Rather evidence, any evidence, is a matter of degree and also carries a subjective and value-based component. The criteria that make for convincing evidence on the part of one person are not necessarily the same as the criteria for another person. Precluding obvious fraud and the like, there is no single magical scientific way to "objectively" determine the ultimate value of any particular alleged evidence. Science, any science, is not quite as "objective" as some would assert.

Quantum Philosophy and the Ancient Mystery School

TODAY'S NEW SCIENCE PHILOSOPHY—OLD OR NEW?

EDWARD F. MALKOWSKI

Responsible for the microelectronic technology that brought us the cell phone, the computer, and the Internet, quantum physics has proven to be history's most successful scientific theory. Quantum physics is also the source of a new understanding of the world around us.

Although the founding principles of quantum physics were developed in the 1920s and 1930s, it wasn't until the 1970s that its influence seeped into our cultural worldview. In 1975, with the endorsement from one of quantum physics' founding theorists, Nobel laureate Werner Heisenberg, Fritjof Capra explored the similarities between quantum physics and the Eastern mystical tradition in *The Tao of Physics*. Another landmark book was published four years later, Gary Zukav's *The Dancing Wu Li Masters: An Overview of the New Physics*. With these books, and many others that followed, a new worldview started to emerge, embracing the interconnectedness between man and nature.

Everything is connected through a universal field of virtual particles, and we are all part of a single living system. What this new worldview suggests is that physical form as biological consciousness is a local expression of a universal phenomenon commonly referred to as "consciousness." The cycle of life and the evolution of form are natural processes that create a framework for experience where consciousness is a fundamental aspect of reality just as much as the spatial dimensions. Consciousness, once thought to be only the product of brain chemistry, is now viewed as the eternal driving force for all that exists, that manifests itself through physical form in order to experience.

Although Mind is very much an enigmatic and highly debatable concept, this new worldview also suggests that the individual's mind is a process of this universal drive to experience, as opposed to being a separate entity. Erwin Schrödinger, another founding theorist of quantum physics, and also a Nobel laureate, views this question of one Mind and many minds as an arithmetical problem. For Schrödinger, our perception is scientifically indescribable, because the mind itself is that world picture. Thus, the individual mind is identical to the whole "Mind," and therefore cannot be contained in it as a part of it. This creates a problem, because there are a multitude of individuals experiencing consciousness, but there is only one world.

One answer to this paradox is that each of us experiences a unique world, which Schrödinger summarily dismisses. There is only one other alternative. The multiplicity of minds is only apparent; in truth there is only a single Mind. Such a concept brings with it complex ramifications for the definition and nature of knowledge.

Secret Wisdom—Sacred Science

You might think that since quantum physics is a relatively new branch of science, this burgeoning "new science" philosophy is also new. It is not. These new insights into nature and reality are very old but have been masked by modern attempts to characterize the ancient Egyptian culture and religion as primitive. The concepts of mind and consciousness, as well as reincarnation and evolution, were expressed long ago in what historians have labeled the "ancient mystery school"—what Schwaller de Lubicz termed "sacred science." Although shrouded by the secrecy of the temple and rites of initiation, ancient Egyptian schools taught this secret wisdom through myth and symbolism, an approach that leads to an understanding of the world that is virtually identical to today's new science philosophy.

In fact, the sacred science of the ancient Egyptians, best described as a philosophy of nature's principles, inspired the Hebrews, the Greeks, the Romans, and the Christians, which led to the emergence of what we call Western civilization—but for us, thousands of years later, the founding knowledge of our civilization is all but lost. However, there have always been people who have handed down the secret wisdom and the sacred science of the ancient Egyptians: Kabbalists, Hermeticists, Gnostics, Sufis, Buddhists, and Alchemists. It is secret only in the sense that this wisdom must be understood through esotericism and symbol, and it's sacred only in the sense that scientific investigation inevitably leads to an understanding of Man, Divinity, and a unique knowledge of "Self."

Leaving behind modern biases and looking deep into ancient Egypt's civilization, there can be found a brilliance and understanding that rivals our knowledge today. Their "gods" were of a different order than our Western concept of God. They were not "gods" at all, but principles of nature that represented concepts like digestion and respiration. They also represented intangible qualities found in mankind, such as knowledge and personality. This ancient view of nature has been mistaken as religious and cultlike, but is, in fact, technical and philosophical.

For example, the king's diadem, with its serpent and vulture, symbolized the principles of life and form. The serpent represented the concept of the Source for all that exists and its manifestation as the cosmos; and the vulture, man's spiritual immortality. Like a spirit, the vulture, soaring high in the sky, escapes this world to an existence beyond the bounds of Earth. Thus, the pharaoh's diadem symbolized man's kingship in a cosmic sense and the mystery of life's essence, where the mystery is the reality of Cause and Effect. This mystery, which defines the human experience, is abstract, but operates through the concrete court of three dimensions to create another abstraction—what we experience as consciousness and self-perception.

How the ancient Egyptians developed such a refined philosophy is a mystery in itself. Scholars such as Samuel Mercer, who translated Saqqara's Pyramid Texts during the 1950s, have noted that the tenets of this philosophy appear to have emerged fully-fledged nearly five thousand years ago, without historical precedent. It is ironic that ancient Egypt's technical capabilities—so ambitious, so precise—also appear to have emerged, fully-fledged, without precedent. Although we shouldn't be surprised, since the development of a sophisticated philosophy does not occur without sophisticated technology.

Such insight into ancient Egypt's earliest traditions moistens the seeds of doubt for history's linear model of man and civilization—particularly so when today's emerging "new science" philosophy parallels concepts described long ago in Ramses' Temple of Amun-Mut-Khonsu, so meticulously described by Schwaller de Lubicz in his two volume work *The Temple of Man.*

In 1937, alchemist and Hermetic philosopher René A. Schwaller de Lubicz was drawn to Egypt by an inscription at the tomb of Ramses where the pharaoh was depicted as the side of a right (3:4:5) triangle. For Schwaller de Lubicz, this meant that the ancient Egyptians understood geometry's Pythagorean theorem long before Pythagoras was born. Intrigued, he moved to Luxor and studied ancient Egypt's art and architecture for thirteen years, concluding that the temple architecture was a deliberate exercise in proportion. The temple, in its detail, described the nature of man as a science, a philosophy that Schwaller de Lubicz termed the "Anthropocosm" or "Man Cosmos."

Philosophy of the Anthropocosm

The question of who we are and why we are here will likely remain the ultimate mystery. Intuitively, however, this mystery can be understood upon the realization that our existence as a conscious biological form can be traced to cosmic events, and that the prerequisites for our existence can be traced to a universal state. Our Earth is dependent on the sun and the solar system in which it is gravitationally trapped; which is dependent on the Milky Way galaxy, in which it is gravitationally trapped; which is also held in place by other forces including, but not limited to, our neighboring galaxies. Any interruption in this line of cosmic dependency would likely result in the cessation of our existence. Thus, it can be said that the cosmos is the true nature of man, and form is the sole means of its expression.

Although it seems as though we are insignificantly small compared to the rest of the universe, there is a single truth to our existence that cannot be denied and that lends credence to the abstract nature of man—the reality of the observer. We observe and perceive an ordered, yet dynamic, arrangement of energy that we naturally translate into sight, sound, smell, taste, and touch. To take away the measures of this reality means reality's destruction, which suggests that the universe was never concrete in the first place. We only perceive that it is. Therefore, like Plato in his cave, we can conclude that the concreteness and form in which we live are really only the knowledge of such things. Einstein agreed implicitly in one of his famous statements: "Reality is an illusion, albeit a very persistent one."

The most interesting question is, where does our ability to observe and perceive come from? According to physicists, it comes from an event called "state vector collapse" where all possible states of the system (the universe) collapse into a single observed state.

During the 1920s, while Heisenberg and Bohr were further developing quantum theory, they realized that a new viewpoint had to be created to achieve a proper understanding of the quantum world. The classical view of a discrete world would simply not work. To accomplish this, they embraced the idea that the world is fundamentally not a collection of discrete objects, but an indistinct, unified world of energy where, at times, discrete objects are perceived. Heisenberg developed his wave matrix theory, and Schrödinger his wave mechanics, to explain their insights. Although slightly different in their approach, these two theories offered a more accurate description of the atomic structure than did classical physics.

What their theories state is that all matter exists as a wave structure that we cannot directly see. What we do see is the localization of the wave structure with its release of energy, which is a simple way to explain state vector collapse. The energy released is what physicists call a photon (a particle of light). We perceive the released energy as a particle, even though it is really a wave. This occurs for us because that is how the human brain works.

Without state vector collapse there would be no perception of separation, no form to experience and, consequently, no expression. The cosmos would remain in an undefined state of absoluteness, a potential of all cosmic possibilities.

All matter that makes up the cosmos is actually configured energy that now exists as a result of stellar nucleosynthesis and supernova. Carbon, nitrogen, oxygen, and other heavy elements—the building blocks of life—were created as a result of large stars collapsing under their own weight and then exploding with tremendous heat, spreading newly created elements into empty space to form interstellar clouds. New research suggests that even amino acids, important for protein synthesis, were formed in interstellar clouds. Thus, scientists argue that since the elements that make up our bodies are the results of a cosmic process, then we are made from stardust and are literally children of the stars.

The Big Bang origin of the universe has been the model of choice for cosmologists for many decades now, but it has always been a scientific paradox. Our known laws of physics are not valid until after the moment of the Big Bang. So, how do we arrive at a universe that we experience, which sprang from nothing? Perhaps the Big Bang is only a perspective to explain the current body of scientific data, and does not accurately represent actual events. As is all of nature, perhaps the universe is cyclical and oscillates between the never-ending destruction and creation of galaxies. No one really knows.

However, what we do know and can be certain of is our conscious experience. It is the one thing all six billion of us can agree upon, and the key to understanding nature. According to the Anthropocosm theory, consciousness creates a venue in order to experience, and does so through the unique quantification of its qualities. This cosmic and anthropic "new science" understanding of man puts forth the same principles that were built into the architecture of Luxor's Temple of Amun-Mut-Khonsu.

The temple was not about the piety of a man, but our solar legacy as the philosophical "Divine Man" portrayed in the great statues of Ramses—the birth of the sun. The temple was (and is) a form of communication, a lesson, and at its core its builders' philosophy is carved in stone. Amun, Mut, and Khonsu were not "gods" in the Western religious sense, but principles that form and explain the nature of mankind as coherently as such an abstract subject can be explained.

The definition of man and the story of the human experience were built into the temple architecture. Physically, the temple describes the structure of man, from the importance of the femur in the creation of blood cells, to the role of the pineal gland in the brain. Spiritually, the temple conveys life's cosmic drama and man's spiritual immortality. Amun was the "Hidden One" or the "Invisible One," best described today as the Western concept of God, omnipotent and omnipresent, or, from a scientific viewpoint, the energy field that pervades all that exists. From the ancient Egyptian point of view, Amun was self-created, the creative power and source for all that exists. Mut, which means "mother," was Amun's cosmic wife and the mother of the "Son" Khonsu, who represented the king.

However, the kingship of Khonsu was not a physical kingship but refers to a cosmic (or spiritual) ruler made flesh through the principles of nature. Thus, Khonsu the King represents the essence of mankind—the archetypal "Man"—and essence of all who ever lived, is alive now, and will live in the future. Khonsu, by being associated with Re and Thoth, represented the essence of life's energy and man's wisdom and knowledge, where mankind is a consequence of the universe's evolution culminating in the physical endowment of the universe's self-perception. In myth, Khonsu was a lover of games, but was also the principle of healing, conception, and childbirth. Literally, he was "the king's placenta."

Just as the ancient Uroboros—the circular serpent biting its tail—symbolizes cyclicality, through our modern scientific endeavors we have come full circle in understanding ourselves. No one knows for sure in what culture or at what time the Uroboros was first fashioned as a symbol, but it is one of mankind's most ancient ones.

Plato tells us in the *Timaeus* that since nothing outside of the serpent existed, it was self-sufficient. Movement was right for his spherical structure, so he was made to move in a circular manner. Thus, as a result of his own limitations, he revolves in a circle, and from this motion the universe was created. From Egypt's Ptolemaic period, the artist who drew the "Chrysopoeia [gold making] of Cleopatra" wrote within the circular serpent: The All Is One. Thus, the serpent is the ancient Egyptian symbol depicting self-creation and the source of life: "It slays, weds, and impregnates itself," writes Erich Neumann in *The Origin and History of Consciousness*. "It is man and woman, beginning and conceiving, devouring and giving birth, active and passive, above and below, at once."

For the ancient Egyptians, the Uroboros—the serpent—represents the creative principle of the cosmos, as well as the cosmos itself. Since the serpent's form is singular, without appendages, but has a forked tongue and forked penis, its form is an apt symbol of creation's initial movement from an undifferentiated state into a world of multiplicity; a movement from one to two. Schwaller refers to this as the "Primordial Scission."

The Uroboros, however, is not just an ancient mythical symbol, nor is it the fabricated imagery of the primitive mind. Rather, it is man's identification with the seamless, eternal state of oneness whose essence is a deep memory of an origin that words cannot explain and has to be understood through esotericism. As such, the Uroboros' esotericism is as valid today as it was at the dawn of man.

The Spiritual Technology of Ancient Egypt

The Western worldview has a long history of separating the physical from the conceptual; the scientific from the religious. So together, spirituality and technology appear contradictory. This contradiction, however, is based on a naïve and exoteric view of "spirit" and "technology."

Spirit is not some immeasurable, metaphysical thing. Rather, Spirit is the driving force behind the human experience, the quest for knowledge, and the building power of civilization that can be measured by achievement. Technology is mankind's application of knowledge into industry that provides for the civilization's well being. Technology, which is the application of science into civil practicalities, is also the building power of civilization.

Even though technology's final product is most evident, it is the spirit of man that turns ideas into concepts, and concepts into knowledge, which through engineering brilliance, turns science into technology and makes life more efficient and comfortable. Every product ever made began with someone's inspiration and creativity. So, spirit and technology are really different aspects of the same human endeavor.

The desire to know inspires us, and the ever-increasing level of knowledge and technology has allowed us to reach new levels in understanding our state of existence—but what might have inspired the ancient Egyptians? Schwaller de Lubicz believed that ancient Egypt was the legacy of a technical civilization of which there is no history or knowledge in today's world, a civilization for which spirit and technology were integrated into a worldview that embraced life's mystery. For me, it is this technical and spiritual legacy that is so evident in the art and culture of ancient Egypt.

The spiritual technology of ancient Egypt expounds upon the works of Schwaller de Lubicz and tells the untold story behind the birth of the Western religious tradition. The Egyptian Mysteries, as they were called, inspired greatness in men who instilled the concept of the Anthropocosm into our own sacred literature, and it is the same philosophical understanding of nature that is at the forefront of today's "new science"; whether symbolized by the Uroboros or Schrödinger's wave equation, human consciousness exists as a local manifestation of a self-perceiving universe.

THREE

The Egypt Code

IS THE KEY TO EGYPT'S PAST REFLECTED IN THE STARS ABOVE GIZA?

ROBERT BAUVAL

What are Egypt's Old Kingdom pyramids for? What possible purpose could they have had? Why do they have low protracted tunnels, and long narrow shafts that seem to lead nowhere, and corridors, galleries, and chambers that are often stark and empty? Why were the pyramids astronomically aligned to the stars? Why are they scattered in clusters along a forty-kilometer strip of desert? And, more intriguingly, why are some devoid of texts while others have their walls fully covered with inscriptions that speak of the sun and stars, and of a strange religious cosmology in a celestial landscape that is reminiscent of Egypt itself?

Until very recently, the standard theory offered by Egyptologists was that the pyramids were tombs, large sepulchers created principally to house the body of dead kings. As for their elaborate internal systems of tunnels, shafts, corridors, and chambers, these were intended mainly to confuse and outsmart tomb robbers, while their astronomical alignments were either meaningless or just a fluke. Amazingly, such views went mostly unchallenged for nearly two centuries, this in spite of the maddening detail that no bodies of kings (not a skeleton or skull or even a bone splinter) were ever found inside a pyramid or, for that matter, outside it.

More maddening still, no one had an explanation why, if they were "tombs," these pyramids were not placed into a single well-defined cemetery but instead were scattered in small clusters in a vast desert plain west of the strange, River Nile–like volcanic islands in a sea of sand. Yet, oddly enough, the clues that suggested a much higher purpose than just "tombs" were plentiful, and always there for all to see and evaluate—and these clues screamed of a connection with the stars. For example:

1. The base of each pyramid was aligned to the astronomical directions using star alignments.
2. The largest of the pyramids contained "air shafts" oriented toward important star systems such as Orion, Sirius, and the circumpolar constellations (viz. the Pyramid of Khufu at Giza).
3. Pyramids were given "star" names or names implicit of stars ("The Pyramid of Djedefre is a Sehedu star"; "Nebka is a star"; "Horus is the star at the head of the sky," and so forth).
4. Pyramids had ceilings of chambers decorated with five-pointed stars (viz. the Step Pyramid and Fifth and Sixth Dynasty pyramids at Saqqara).
5. Pyramids contained writings carved on the inside walls that spoke of a star religion and the destiny of a king in a starry world called the Duat, which contained Orion and other constellations (viz. the Fifth and Sixth Dynasty pyramids at Saqqara).

It is therefore somewhat odd, not to say perverse, that with so many "stellar" connections, there has not been a single Egyptologist who was compelled enough to consider a stellar "function" for the pyramids. And because this important matter was left unbridled for so long, it was not surprising that untrained researchers, dilettantes, cranks, and charlatans dished out theories that ranged from the derisory to the completely insane. Pyramids were built by the lost civilization of Atlantis; they were built by a lost technology using levitation; they were power plants; they were electromagnetic receivers for interstellar communications; they were built by aliens; they were built by the Jews while in captivity in Egypt; the Great Pyramid was designed to contain detailed information of world history and future in every inch of its plan; it was a Bible in stone.

So when I burst on the scene in 1994 with my first book, *The Orion Mystery,* showing that the pattern of the three Giza pyramids and their position relative to the Nile mirrored the pattern of the three stars of Orion's Belt and their position relative to the Milky Way, the subject was so much soiled and degraded that any new theory that mentioned the "stars" or "astronomy" was immediately met with a barrage of academic indifference (at best), or vociferous opposition. The reaction was even more violent because my theory had received the (albeit cautious) backing of one of the world's most eminent and most respected Egyptologists, Sir I. E. S. Edwards, who had gallantly and boldly stuck his neck out on my behalf by appearing on the 1994 BBC2 documentary *The Great Pyramid—Gateway to the Stars,* in support of some of my ideas.

This brought him the wrath of his peers, but it nonetheless twisted their arms and forced some to grudgingly review my theory. However, in the years that followed, and especially after Edwards' death in 1996, I was derided and pilloried by a cabal of Egyptologists and other "experts" seemingly determined to "debunk" the Orion Correlation Theory, as my hypothesis was now being called. All this academic onslaught was most daunting and distressing, but I firmly held my ground, for I knew that I had not only generated massive interest and support in the general public and the international media but that the theory I had proposed neatly dovetailed into the context of Egypt's Pyramid Age and provided the "missing link" to an otherwise baffling mystery. Even the most entrenched skeptic could not easily dismiss the Orion-Giza correlation as "coincidence."

Fourteen long years have now passed since the publication of *The Orion Mystery*. Meanwhile, the book has been published in more than twenty languages and there have been dozens of television documentaries fully or partially-based on the Orion Correlation Theory (viz. Britain's BBC2 and Channel 4; America's ABC, NBC, and FOX; Europe and America's Discovery Channel and History Channel; Italy's RAI 3; Germany's ZDF and ARD; France's ARTE and TF3; South Africa's SABC and M-net TV; Holland's AVRO TV; Australia's Channel 7; Egypt's NILE-TV and many more other channels in the Far East and Middle East).

Two more documentaries were aired with National Geographic television, both entitled *Unsolved Mysteries of the Pyramids* (where my theories were critically reviewed), and another, *Egypt Decoded,* made for Italy's RAI 2 and Holland's AVRO fully based on *The Egypt Code*. Slowly but surely, the Orion Correlation Theory has crept like a thief in the night into mainstream Egyptology and the new discipline of archeoastronomy. And even though it is given much lip and criticism, it is very obvious that it has touched the proverbial nerve of academia.

To be fair, not all academics were prone to dismiss *The Orion Mystery*. Some very eminent Egyptologists, such as Dr. Jaromir Malek of the Griffith Institute and the American Egyptologist Dr. Ed Meltzer kept an open mind just as the late Sir I. E. S. Edwards had. More refreshingly, the theory received cautious support from the astronomical community, particularly from Professor Archie Roy of Glasgow University, Professor Mary Brück of Edinburgh University, Professor Giulio Magli of Politecnico di Milano, Professor Percy Seymour of Plymouth University, and Professor Chandra Wikramasinghe of Cardiff University. And even though these high-ranking astronomers maintained a healthy skepticism, they nonetheless found the theory intriguing and deserving of careful consideration and further research.

Also in the course of the years, a crack began to appear in the Egyptological academic armor when Dr. Jaromir Malek (who had reviewed my theory in 1994 in the Oxford journal *Discussions in Egyptology*) declared himself favorable to the possibility that the apparent illogical scattering of pyramids in the Memphite necropolis (a 24.8 mile [40 kilometer] long desert strip west of the Nile near Cairo) may, after all, have had more to do with "religious, astronomical, or similar" considerations than with purely practical considerations such as the topography and geology of the land. Similar views began to be heard in Egyptology, especially from the American Egyptologist Mark Lehner, the Czech Egyptologist Miroslav Verner, and the British Egyptologist David Jeffreys.

It was, however, the archaeoastronomer Anthony Aveni, a professor of astronomy and anthropology at Colgate University, who, in my view, would come the closest in providing an overall picture of what may have been in the minds of the ancient architects who designed and planned such mysterious monuments (not only in Egypt but in other parts of the ancient world). He wrote that, "In order to understand what ancient people thought about the world around them, we must begin by witnessing phenomena through their eyes. A knowledge of each particular culture is necessary, but learning what the sky contains and how each entity moves is also indispensable . . . strange but true: Whole cities, kingdoms, and empires were founded based on observations and interpretations of natural events that pass undetected under our noses and above our heads."

Dr. Aveni was referring to the Maya and Inca civilizations when he made the above statement—but he may as well have been talking about Egypt's Old Kingdom, for I am now even more convinced that such a statement holds truer for the sacred cities, pyramids, and temples built by the ancient Egyptians along the 621.4 miles (1,000 kilometers) of the Nile Valley during their three thousand years of civilization. And this, in a nutshell, is what I set out to prove in *The Egypt Code*.

By the year 2000, I was ready to put the findings of my investigation into book form. To this end I presented a synopsis to my editor at Random House in London, who promptly commissioned the project. By early 2004, I had a first draft ready. The final draft, however, was completed in Egypt. In February 2005, I rented a fully furnished apartment with a direct view over the Giza pyramids. Being here gave me the unique opportunity to refine the book with a hands-on approach to the pyramids in Lower Egypt and the great temples of Upper Egypt, and to verify and test the various ideas of my thesis. Imbued with the enchantment and magic of these ancient sites I have, I believe, succeeded, in more ways than one, in bringing the sky-ground correlation theory I started two decades ago to its natural conclusion.

In *The Egypt Code,* I have made use of primary sources whenever available, and relied only on scholarly research published in peer-reviewed journals or in textbooks by renowned Egyptologists and other scholars. My readers should expect no less from me. Culling my data from all of these sources, I have come to this conclusion: The ancient Egyptian theocracy was regulated by a cosmic order called Ma'at, which was none other than the order of the sky: the observable, precise, and predictable cycles of the sun, the moon, and the stars. I have also concluded that this Cosmic Order was fervently believed to influence the material world below, especially the all-important annual flooding of the Nile, for nothing more fascinated, awed, and *frightened* the ancient Egyptians than the Nile's flood, which began in late June and ended in late September. This was the annual miracle that rejuvenated the crops and all other life in Egypt, or brought famine and pestilence if the waters failed to rise.

This double-edged sword that hung perpetually over Egypt compelled the Nile dwellers to seek magical means to ensure a good flood. Early in their development they came to observe that the stars of Orion and Sirius would disappear underneath the western horizon after sunset in late March and remain for a protracted period (about three months) in the "underworld" before reemerging on the eastern horizon at dawn in late June *just when the waters of the Nile began to rise.* During this crucial period of the stars' sojourn in the "underworld," the astronomer-priests also noted that the sun traveled from a point on the ecliptic just below the bright cluster of the Pleiades (marking the vernal point) to a spot farther along the ecliptic just below the chest of the celestial lion, Leo (marking the summer solstice), that bracketed the constellation of Orion and Sirius.

The idea began to enter their minds that when the Sun God journeyed through that special part of the sky—the Duat, as it was called—he performed a magical ritual, a sort of "stations of the cross," that would bring about the "rebirth" of the stars as well as the "rebirth" of the Nile when, in late June, the star Sirius would reappear at dawn on the eastern horizon. This event happened to also fall on the day of the summer solstice, when the sun would reach its maximal northerly declination, and was for good reason taken as New Year's Day and called (among other things) the "Birth of Ra," the Sun God.

A mythology and sky-religion developed around this cosmic and Nilotic theme and, more intriguingly, an ambitious plan was gradually hatched around 2800 BCE to "bring down," in the literal sense, the Cosmic Order so that the pharaoh, the son of Ra on Earth, could undertake the same magical journey in an earthly Duat and thus secure for Egypt a "good" flood (recall the Hermetic dictum: *As above, so below*). To this end, a massive pan-generational project was put into action that would involve building clusters of "star"-pyramids at predetermined sites to represent Orion and the Pleiades, as well as the building of great "sun"-temples set on both sides of the Nile to define the part of the ecliptic along which the Sun God traveled through the Duat from vernal equinox to summer solstice, set on both sides of the Milky Way.

My new theory doesn't stop there—I was also able to determine that the slow cyclical changes witnessed in the sky landscape caused by precession and by the peculiarities of the Egyptian civil calendar over the three thousand years of the pharaonic civilization are reflected in the changes witnessed on the ground all along the 621.4 miles (1,000 kilometers) of the Nile Valley in the evolution of temples throughout the same three thousand years. In other words, *The Egypt Code* proposes, no less, to prove that there exists a sort of "cosmic Egypt" ghosted in the geography of the Nile Valley stretching from north to south that was once literally regulated and administered by astronomer-priests headed by a sun-king; that this cosmic Egypt lasted for over three millennia, and that it can still be discerned in the layout of the pyramids and temples that remain today.

My thesis is entirely verifiable, testable, and ultimately falsifiable if need be. Indeed, I happily welcome Egyptologists and other scholars and researchers in the field of Egyptian archaeology and history to step up to the debate.

FOUR

Alternative History and Esoteric Philosophy

TWO PORTALS INTO THE SAME WORLD?

MARK BOOTH

Something that often surprises me about the opponents of alternative history is how readily they stoop to intellectual dishonesty. Typically the "scientifically correct," as I like to think of them, present themselves as high-minded defenders of intellectual rigor. They then go right ahead and attribute to others claims that they have never made, just so they can rubbish them!

The 1999 BBC2 *Horizon* documentary about Graham Hancock and Robert Bauval, "Atlantis Reborn," was a shining example of this—cut, dried, and well documented in the adjudication of the Broadcasting Standards Commission. Maybe it's a kind of tribute? If the scientifically correct were sure of their ground, perhaps they wouldn't feel the need to behave so shiftily?

Yet they do, and more's the pity on several deep and important levels, as alternative history often touches on what the existentialist theologian Paul Tillich called "the ultimate questions"—where we come from, who we are, and what the meaning of life might be. Of course, ideally, these questions should inspire a wholehearted desire to discover the truth. We should be passionately interested, yet scrupulously disinterested, setting aside all partisan affiliations, even the desire to be right, because the answers we give to those ultimate questions determines the very way we choose to live our lives.

Exactly how *does* alternative history bear on the ultimate questions? I think that this is best explained using an example that is central to alternative history. If the Sphinx dates back deep into what's conventionally called the Stone Age—in other words if it is many thousands of years older than conventional, academic history allows—then it follows that we are not who we thought we were. Our history has different patterns than the accepted ones.

This question of the age of the Sphinx is also an example of a curious feature of the human condition as a whole, a feature that is quite remarkable, but often overlooked: *When it comes to issues like these, we find ourselves dealing with minute fragments of evidence that admit of many different interpretations, sometimes even contradictory ones.*

It seems to me that when it comes to the great questions of history, and to the great questions of life and death that are tied up with them, the evidence is often not so overwhelming that it imposes an answer on us. *We often have great latitude when we choose what to believe.*

Perhaps we can then choose what we want to believe?

Important, then, to be aware of which part of ourselves is doing the choosing, that we do not choose unconsciously, but instead bring our full intellect to bear. Is it the part that really wants to know the truth that is doing the choosing? Or is it the partisan, egotistical side that wants to be right or to be on the winning side?

The fact that we are in a position to consider the ultimate questions in a relaxed and tolerant way, and without trying to tear each other's throats out, the fact that a wonderful forum like Graham Hancock's website exists, is at least in part due to the work of secret societies—in particular, the secret societies that lay behind the Royal Society, and therefore the great scientific and industrial revolutions of the eighteenth and nineteenth centuries. These secret societies created protected spaces (sometimes called lodges) where freethinking, disinterested intellectual enquiry could take place.

In these spaces, people like Isaac Newton, Robert Boyle, Robert Hooke, and Bruce Harvey were not only able to discover and define gravity, formulate the law of thermodynamics that paved the way for the internal combustion engine, invent the microscope, and uncover the workings of the circulatory system, they were also able to pursue their interest in alchemy and other arcane subjects. When an outsider questioned Newton about his interest in astrology, he is reported to have replied, "Sir, I have studied it, you have not." Newton also believed that we live in a world dense with secret codes—in the laws of nature, in books like the Bible, and in ancient monuments like the ones on the Giza Plateau. They were put there, he believed, to help draw our intelligence out of us.

The initiates of the secret societies had realized that you get two very different sets of results if you look at the world as objectively as possibly and then, on other occasions, as subjectively as possible. This realization brought great material benefits to the world, but it also opened up many strange realms of thought. . . .

It was brooding on these sorts of things, especially the dates of the monuments on the Giza Plateau, that led me to think that I might have a contribution to make as a writer. In my day job, I was editing and publishing (at different times) not only Graham and Robert, but also Robert Lomas and Christopher Knight, Robert Temple, David Rohl, and Michael Baigent. In my spare time, I had also developed an interest in esoteric philosophy, in Theosophy (with a big as well as a small "t"), in the Rosicrucians, and in their modern representatives, the Anthroposophists. I used to delight in finding obscure and weird old books about the esoteric and mystical in secondhand bookshops—for example, the works of magi like Paracelsus and Jacob Böhme.

It struck me that, although as far as I knew none of these alternative historians were—at that stage, at any rate—much interested in esoteric philosophy, many of their discoveries were confirming its tenets, regarding, for example, the claims that the Sphinx and the Great Pyramid are much older than conventional history allows, or related claims regarding the historical reality of Atlantis and the Flood.

If new evidence was being unearthed that suggested that extremely important traditions like these have some basis in historical reality, the question naturally arises: what other traditional esoteric information about history might also be true? So I planned to try to weave together, into one narrative, historical lore from different esoteric traditions from around the world. To achieve one narrative thread, I decided to focus on what they had in common—discarding what are sometimes called "cultural accidentals"—and also to focus, where possible, on traditions that chime in with the latest discoveries of alternative history.

I very quickly realized that if this history was to be in one volume, rather than spreading across many, many volumes, it could not incorporate debate as to whether its claims—the arguments for and against the Sphinx being some twelve thousand years old, for example—are true. This would have to be a "take it or leave it" history. If readers wanted to follow up on the pros and cons of these debates, they'd have to turn to the works referenced at the back. (In the case of the Sphinx, these would be books by Graham and Robert, John Anthony West, Robert Temple, and Schwaller de Lubicz.)

After a while, I began to formulate a theory on what these esoteric traditions had in common: they all describe the ways that the supernatural works in the world. Gods, angels, and spirits may have different names in different places and at different times, but, according to secret teachings everywhere, the patterns they help make, and the shapes they give to our lives are the same. Therefore, my resulting book, *The Secret History of the World,* describes patterns that wouldn't be there if materialistic science accounted for everything.

These patterns in history are perhaps deeper than the laws of economics, the effects of climate change, and the conventional, materialistic view of politics that interests modern, academic historians. In other words, I show history operating according to what I call (after the poet Rainer Maria Rilke) the Deeper Laws. Then, at the end of the book I invite readers to look at their own lives to see if they can't find these same Deeper Laws operating there.

In the book, I tried to weave together different mystical traditions about our beginnings and endings and great turning points in between into one epic imaginative vision. My aim was to see if this imaginative vision formed a coherent, cogent whole that might be set against the scientific materialist one. I couldn't think of anyone who had tried to do this since Milton, and he had done it in very different circumstances, when scientific materialism was beginning to roll back the idealism that had been the universal philosophy up to that point. Madly, I asked myself if it were possible to create an imaginative vision that would be a sort of mirror image to Milton's—written at a time when scientific materialism seems to many to be beginning to fray and look a bit thin at the edges.

The world is a much more mysterious place than we have been brought up to believe. There are other ways of knowing than the one we have been taught. The priests and artists of the Egyptian and Hindu temples knew of and understood the function of the pineal gland thousands of years before it was "discovered" by German and English anatomists (more or less simultaneously) in 1866.* Robert Temple has shown that the Egyptian priests knew that Sirius is a three-star system, something only confirmed by French astronomers using radio telescopes in the second half of the twentieth century.

According to Rudolf Steiner—the founder of Anthroposophy—knowledge of the evolution of the species from marine life to amphibian to land animal to anatomically modern human was encoded thousands of years ago in the imagery of the constellations. Jonathan Swift was deeply immersed in esoteric philosophy. In *Gulliver's Travels,* he predicted the existence and orbital periods of the moons of Mars. A hundred years later, when astronomers first observed these moons using the latest telescopes, they named them Phobos and Deimos—fear and terror—so awestruck were they by Swift's evident supernatural powers.

How did these guys know? Where did their "powers" come from? Is there something going on we don't fully understand?

Again, the intellectual establishment would like us to see a very clear distinction between modern scientific endeavor and the occult, seen as primitive and superstitious—but as I demonstrate, the line is not always clear.

We've already touched on Newton, Boyle, Hooke, and Harvey. Newton's rival in devising the calculus, Gottfried Leibniz, made his advances while studying the Kabbalah. Paracelsus, the great Swiss magus, a great practical alchemist sometimes credited with inventing the principles that lie behind homeopathy, has also been called the father of modern experimental medicine. Emanuel Swedenborg, the most famous psychic and esoteric teacher of his day, also discovered the cerebral cortex and engineered the largest dry dock in the world. Perhaps significantly, Charles Darwin formulated his

---

The Parable of the Beast, by John Bleibtreu, published by Paladin in 1976, chapter 2.

theory of evolution shortly after his friend Friedrich Max Müller had made the first translations into a European language. Sigmund Freud was very interested in the Kabbalah as a boy, and his model of the mind—super ego, ego, and id—can be seen as a materialistic version of the Kabbalistic one.

Carl Jung based his account of what he called the Seven Great Archetypes of the Collective Unconscious on the spirits of the planets as they have always been understood in esoteric lore. Thomas Edison, the inventor of the phonograph (and so, in a sense, the godfather of all recorded sound) and Alexander Graham Bell, the inventor of the telephone, made their discoveries while researching the spirit worlds. Edison tried to make a radio that would tune into the spirit worlds. Even television was invented as a result of scientists trying to capture psychic influences on gases fluctuating in front of a cathode ray tube.

Scientists from Newton to Tesla to Einstein have also talked about their great discoveries coming to them suddenly in a dream or in a vision. Einstein even compared the process to "the three degrees of initiation" of the ancients. I was fascinated to hear from Graham Hancock that Francis Crick cracked the code of the double helix and launched modern genetic medicine while he was on an LSD trip.

The reality is that we all perceive things in altered states of consciousness—in dreams, visions, prayerful states, meditations, perhaps under the influence of drugs, and when we're struck by hunches and premonitions. The key question is: *Are any of the things we perceive in these altered states real, or are they all delusions?* In other words: *Do altered states bring other ways of knowing?*

Only a fool would deny that science has brought untold benefits to the world. (And I don't mean "fool" in an interesting esoteric way, I mean just plain stupid.) Science has made our lives safer, easier, more comfortable, longer, and given us more spare time for enjoying ourselves, for art and speculation. However, these benefits have come at a cost.

We have been enabled to make scientific leaps forward because we have been so focused on the material world. We have constructed a practical, commonsensical form of consciousness for ourselves that is great for tying our shoelaces in the morning, navigating our way around our mechanical world in our cars and planes, and fending for ourselves and our families.

Yet other, more subjective forms of consciousness, other ways of knowing, have been squeezed out, devalued, and discarded. Esoteric philosophy is the systematic cultivation of these other ways of knowing. It is also the richest tradition of thinking about "the ultimate questions." The more I researched my history, the more I discovered just how many of the great men and women of history were steeped in it. I began to wonder if it might even be the case that every intelligent person checks it out at some time or other.

In the late nineteenth century, writing of the totality of reality as consisting of "the playground of numberless Universes incessantly manifesting and disappearing," Madame Blavatsky prophesied some of the leaps forward that science would make in the twentieth century. Although many leading scientists in the twenty-first century postulate possible universes and multiple dimensions, I suspect most of us live—in terms of our everyday, unconsidered worldview—in a solid, commonsense sort of universe with just one dimension. There, causality is still a matter of atom knocking against atom like billiard balls—with entirely predictable results.

This leads, I believe, to two closely related and very common logical fallacies. These fallacies are typical of scientific correctness. One is to believe that if an event can be shown to have one cause, no other causes of it are possible. So an example relevant to my work would be to believe that because thunder and lightning can now be explained in terms of clashing blocks of air and electricity, this proves that they cannot be the result of the anger of a god.

Another example that relates to a common present-day belief would be that if you could capture a pigeon shedding a feather on CCTV just before you walked into view, then you would know that the pigeon is the cause of your sighting of the feather, not that an angel is trying to send you a message.

The reality we are more likely to recognize if we don't have an antispiritual ax to grind is that we commonly accept that simple events have infinite networks of causes behind them, and so that, in these examples, there is nothing contrary to logic about supposing that both the mechanical and spiritual explanations may be true at the same time.

One might call this fallacy—with a little ironic nod to the law of excluded middle—the "law of excluded cause."

Closely related to this fallacy is another one to do with the interpretation of texts: if it can be shown that, for example, biblical or mythological stories such as the conquest of the Promised Land, the voyage of Jason and the Argonauts, or the siege of Troy are descriptions of astronomical events or alchemical processes, they cannot be true descriptions of historical events.

That simply does not follow. To borrow Alvin Plantinga's tools of analysis, there are many possible universes in which both these interpretations are true.* Indeed, in the esoteric universe of my book, *all* true descriptions of events on this planet are also true descriptions of astronomical events.

One of the aims of my book was to give consciousness a slight nudge away from the billiard ball view to something more multidimensional, more in tune with the reality of interpenetrating realms.

---

*Alvin Plantinga is one of the leading figures in the Anglo-American school of analytic philosophy. He is currently John A. O'Brien Professor of Philosophy at the University of Notre Dame.

Interpenetrating realms are also reflected in the *form* of some esoteric literature. I'm sure that anyone reading this will be familiar with Kabbalistic number mysticism and the idea that there are other texts encoded within the Hebrew scriptures. A Cambridge-based statistician came to me recently with some texts he had derived using "skip codes," which he had then translated from Hebrew into English. These texts were a bit fragmentary, as if the skip code weren't 100 percent accurate, but in subject matter and form they were somewhat like the Psalms.

It seems to me entirely possible that this kind of research will eventually yield entire other books encoded within the books of the Hebrew scriptures—and the question that naturally arises is will these other books have further books encoded in them, and so on to infinitude?

To write books with other books encoded within them would seem to require supernatural intelligence. Yet great initiates, including Rabelais and his contemporary at Montpellier University, Nostradamus, as well as Shakespeare, are sometimes said to have written in what is called the Green Language—a language that reflects different orders of reality in its different layers of meaning.

I leave you with one last possibility—could we all be thinking, speaking, and writing the Green Language all the time? Could the only difference between us and the great initiates be that we do it unconsciously?

A senior publisher I know—a rather down-to-earth and worldly fellow, I don't think he'd mind me saying—started to read *The Secret History of the World,* saying that he wasn't sure if he could take it seriously; wasn't even sure he was *meant* to take it seriously. In fact, he may have used the phrase "apparently barmy!" However, by the end he said that he had started to see life from a slightly different angle, to notice connections. A senior publicist of a rather more spiritual inclination, who began to read it at the same time, started to experience some mild, supernatural phenomena. The last I heard, the two of them were planning a trip together to the British Museum to stand in front of the Lohan—a statue alluded to in my book, which seems to have mysterious properties.

I can't claim my book offers initiation. As I explain, I am not an initiate myself—but feedback like this reminds me how much help I have had writing this book, and that much of the imagery that flows through the narrative was devised (by minds far more intelligent than mine) to work at a below-conscious level.

The Dawkins tendency has pretty much had its own way recently. I believe that nothing will turn the tide of materialism except real, lived spiritual experience. When you've had that, you don't believe—as Jung said, you *know*. And that, as I try to show in my writing, is what the esoteric teaching of the secret societies has always offered.

FIVE

An Open Letter to the Editors of *Archaeology*

TAKING TRADITIONALISTS TO TASK

JOHN ANTHONY WEST

Re: *Atlantis and Beyond*

The Lure of Bogus Archaeology

(Or: "The Anupadeshi Strike Back")

A Special Section in *Archaeology*, May/June 2003

Gentlemen (and scholars),

An acknowledgment:

I am delighted to learn that our 1993 NBC special *Mystery of the Sphinx* (based upon geological investigations carried out by me and my geologist colleague Dr. Robert M. Schoch of Boston University) has been awarded top honors in your "Worst of Television Archaeology" list. The show "Argues that the Sphinx is thousands of years older than currently believed and includes comparisons of the Sphinx with the Face on Mars (since shown to be wholly natural . . .)."

Short of having my books singled out and condemned by an official George W. Bush presidential decree, I could not feel more flattered. I regret only that you have no physical equivalent of an Oscar or Tony to go with it. It sure would look good on my old oak filing cabinet, right next to the Emmy I won for "Best Research" for *Mystery of the Sphinx,* which, by the way, was also one of four nominated for "Best Documentary of 1993." Anyway, I think that extraordinary disparity in public opinion a healthy sign, don't you—proof that democracy is still with us, alive and well, despite appearances to the contrary? Would you be kind enough to send me a parchment diploma, a little printed certificate . . . something, anything to hang on the wall to prove that our show won?

The Special Section

"Why do people so desperately want to believe in Atlantis-style tales?" moans *Archaeology* editor Peter Young, unable to comprehend why people are turned away from the "real" archaeology featured in his magazine. Bogus archaeology expert Garrett Fagan, assistant professor of classics and author of the epic, bestselling, modern day scholarly classic *Bathing in Public in the Roman World* explains,

> There is little doubt that presenting science (and archaeology) on television is a difficult business. The slow pace of change in scientific thinking, the habitual lack of consensus among academics about details, and the often complex nature of the arguments involved place pressures on producers. . . . The unspectacular and painstaking nature of the discipline does not make for particularly spectacular television. For how long will viewers sit through scenes of dirt-sifting through knee-high ruins?

That may sound plausible, but it is claptrap. Interest in archaeology is no more dependent upon sifting through dirt than interest in baseball is dependent upon spring training or bat manufacture. Like baseball fans, archaeology fans revel in the game—which in this case is not dirt-sifting, but uncovering and interpreting the past. It is the significance and the relevance of those discoveries that generate interest. The key word here is "significance."

The audience will sit through plenty of "dirt-sifting" if the stakes are high and valid. Your prizewinning selection for "Worst Television Archaeology" had its obligatory patina of network glitz, but most of that show was devoted to a complex scientific geological argument. The audience was riveted, and still is. It is not that "people want so desperately to believe in Atlantis-style tales," it is that they are smart enough to recognize the comic triviality of your petty discipline. Again, Fagan inadvertently supplies the clue (just about everything Fagan supplies is inadvertent). Those heated arguments over "detail" (the Big Picture is agreed upon by the "experts"; only "details" remain) appeal to nobody but yourselves. The archaeologically uninitiated cannot be made to warm to furious debates over how many asps killed Cleopatra (*Serpent in the Sky*, p. 9)—especially when profound mysteries, self-evident to all acquainted with the problems involved, go unexplored, their very existence left doggedly unsifted by archaeological consensus.

We do *not* know how the pyramids were built, we do *not* know why they were built (there is *no* evidence, none, that the pyramids of Giza and Dahshur ever served as tombs, though other pyramids did. They may have been tombs, but there is *no* evidence that they were—got it? Science is supposed to be based upon evidence, not inference). We do *not* know how the two-hundred-ton blocks of the Sphinx and Valley temples and the paving blocks surrounding the Pyramid of Khafre were moved and put into place, and so on. People are not as stupid or gullible as you think they are. They don't buy your version of the Big Picture. It's as simple as that. Unfortunately, they also aren't very discriminating. They tend not to distinguish between, say, an Erich von Däniken and an R. A. Schwaller de Lubicz—but then, neither do you.

To understand your unsolvable PR and image problems and the public's stubborn refusal to accept your word as Gospel, all you have to do is reflect upon what you wrote in your own special section—and, more important, what you didn't write. *Mystery of the Sphinx* wins your "First Worst Prize," and yet, in all those pages devoted to the wholesale denigration (and often misrepresentation) of the work of Graham Hancock, Robert Bauval, and everyone else who dares challenge the Sacred Archaeological Status Quo, there is no response to the geology—the water-weathering to the Great Sphinx—precisely that which qualifies it for Worst Place honors. Not a word. How odd! But there is a good reason for it.

Because you have no response, that's why. And it is that (so far) irrefutable geology that justifies and legitimizes the entire search for alternatives—from carefully developed and sound mathematical and astronomical theories down to von Däniken and the wilder shores of alien intervention. Until you find a way to disprove the geology, the search itself is neither "bogus" nor "pseudo," though some of the material cited as evidence may well fit those categories. . . . Curious note: If an archaeologist is disproved on some significant detail (say that, given enough is through dirt, it is established that Cleopatra committed suicide using just one asp after all), multiple asp proponents will be called "mistaken" and "wrong," but if those outside the archaeological Vatican make a no more egregious mistake, they are practicing bogus or pseudo science.

To bring *Archaeology* readers up to date on the geology—since developments in this ongoing investigation somehow do not find their way into your pages—here is a brief update. Our geological evidence was presented first at the annual meeting of the Geological Society of America (GSA) in 1991; further compelling evidence was presented at the GSA meeting in 2000, both times with the overwhelming support of attending geologists—and shrieks of outrage from archaeologists and Egyptologists.

Over the intervening years a handful of opposing geologists, most with a stake in academic archaeology or Egyptology, have offered mutually exclusive alternative theories to account for that weathering ranging from demonstrably just plain wrong (K. Lal Gauri) to certifiably inept and inane (James "Wet Sand" Harrell's theory). All have been easily, systematically, and conclusively dismantled and rebutted point by point. Meanwhile, two English geologists, Colin Reader and David Coxill, independent of each other and of ourselves, have studied the matter on site and support the theory (precipitation-induced weathering) unconditionally, categorically necessitating rethinking the dating of the Sphinx, and along with it pretty much everything archaeologists accept as dogma regarding very ancient history.

The actual dating remains a matter of debate, so the extent of that rethinking process cannot be determined at this point with certainty—but that it must be radical is apparent to all but yourselves. Which is one reason why a quarter of a billion people (rough estimate) have seen *Mystery of the Sphinx* and been won over by it. It is often bought by teachers (sometimes with their own money) to show to students from grade school up through college level; it has considerable support among academics and scientists across a spectrum of disciplines not in danger of caving in from the implications of a vastly older Sphinx. (Threatened by evidence as revolutionary as this in their own fields, they'd probably react as you do, but that is not the issue here.)

It is even taken seriously by a handful of credentialed Egyptologists and archaeologists, who mostly keep quiet about it, not wanting to subject themselves to the predictable academic auto-da-fé they know will be their lot if their interest is revealed. It is, however, discussed briefly (but taken seriously) by Egyptologist Edmund Meltzer in his essay on the history of Egyptology in (that encyclopedia of New Age pseudoscholarship) *The Oxford Encyclopedia of Ancient Egypt,* edited by Donald Redford. Closet New Age flakes show up in the strangest places, don't they?

While our geological evidence does not in itself prove the existence of a physical Atlantis (we never said it did), it goes a long way toward proving the existence of *an* "Atlantis," a highly developed civilization capable of moving around two-hundred-ton blocks of stone at a time when civilization is not supposed to have existed at all. In other words, it scuttles the historical context of your entire discipline. Yet there's not one word about the geology in your pages of carefully orchestrated debunkery masquerading as scholarship.

Strange omission! Yet hardly unique. Your fourth-place "Best of Television Archaeology" entry, "'Atlantis Reborn Again' (Systematic Dismantling of Graham Hancock's Proposition About His 'Lost Civilization')" resorts to the same chicanery. In that television equivalent of *Archaeology*'s hatchet job, there is also no mention of the geology of the Sphinx. A long filmed interview with Schoch on the subject was carefully edited down to a brief appearance in which he gives his negative opinion on the underwater Yonaguni formations. (This striking site, with its remarkable geometric angular ledges and walls is believed, by Hancock and others, to be man-made, or at least man-doctored. If so, it would be the "smoking gun" testifying to the "lost civilization" we are looking for. Schoch and I are 99 percent convinced that it is wholly natural—but it is always wise to leave that 1 percent open. As they say in the ads for the New York State Lottery, "Hey! You never know.") So Schoch's opinion was good enough to refute Yonaguni, but not good enough to support the water-weathering to the Sphinx. There's no mention of that.

In a court of law, that's called "withholding evidence," and it's a crime. If academic malpractice were a crime (not a bad idea!), a lot of archaeologists wouldn't be walking around on the loose. Chris Hale, the producer of "Atlantis Reborn Again," largely escaped the consequences of withholding evidence. As unprincipled as you, but less maladroit, he had the wit not to hand out a "First Worst Prize." And it is difficult to prove "intent" when the evidence is totally excluded—but you, in your debunking zeal, could not resist calling attention to the geology by bestowing the prize and then conveniently failing to discuss what qualifies it for the honor.

Presumably you thought no one would notice? Now they will. Thanks to the Internet hundreds of thousands, maybe millions will notice.

The Reason Why

But while this reveals the modus operandi of your scholarship, it does not really account for the attraction to alternative views that motivates the special section. That attraction is actually justified by Fagan, even as he discounts it. "Pseudo-archaeology fans," he sneers, "get attracted to all sorts of odd notions. Their ancient civilizations are better than ours, more peaceful, more spiritually attuned."

Now this is a curious, wholly subjective statement coming from a self-proclaimed scientist supposedly devoted to objective truth, and it merits dissection (or, more accurately, trisection). "Better" is not an "odd notion." It is a judgment call. Me? I would prefer to live in a world without hydrogen bombs and traffic jams, a world where you can drink the water. On the other hand, even I would rather go to a twenty-first century dentist than a Twenty-First Dynasty dentist. On balance, I'd say ancient was "better," but it depends entirely upon how one's values are weighted. Fagan has every right to disagree—which, in turn, I might call an "odd notion."

"More peaceful?" Well, since the world has never been less peaceful than it is at present, this does not seem such an "odd notion" either, especially if you go back far enough. Old Kingdom Egypt was most assuredly more peaceful than anything around today. That is demonstrable.

"More spiritually attuned?" Here, inadvertently as always, Fagan has stumbled upon the core of the matter. That ancient civilizations were more spiritually inclined and directed (I'll get to "attuned") is undeniable. A civilization may be judged infallibly by what it does with its collective creative energy. (You do not have to be a Christian or even religious to recognize the truth of "Ye shall know them by their fruits"—Matthew 7:16.)

We put the bulk of our creative energies into shopping malls, weapons of mass destruction, Hollywood and television trash, bobble-head dolls, and Disneyland, with a dollop left over for clever but emotionally bereft science and technology—most of it destructive and/or frivolous, a small percentage of it undeniably beneficial. Egypt (and all other ancient peoples, to a greater or lesser extent) put their creative energies into temples, tombs, and pyramids, all designed to facilitate the quest for immortality. This is a "fact" that should be apparent even to archaeologists—but does it qualify for "spiritual attunement?" I'd say yes. All you have to do is go to Egypt and experience it for yourself. It is self-evident . . . to all but the emotionally defective and spiritually dyslexic. Still, it's a judgment call. It has nothing to do with "science"; not our science, at any rate. Yet that it should qualify as an "odd notion" in the Fagan lexicon is revealing.

To our Church of Progress (materialist, rationalist, Darwinian), "spiritual" is a synonym for "superstitious," and "spiritually attuned" is therefore meaningless. There can be no attunement if there is no spirit. So why use that particular word? Why not grant the ancients their demonstrable ancient preoccupation with superstition, and leave the pseudo-archaeologists to theirs? After all, we are not threatening your biological survival—the only "value" permissible in your one-dimensional Darwinian cosmos (though that, too, is purely subjective; I won't go into that here). Yet it is deemed dangerous. Pseudo-archaeology must be contested, stamped out. Why?

My old Japanese *sensei* put a finger on it. He used to counsel, in his broken but pungent English, *"You want happy in klazy world? No talk moonbeam to blind man; no talk music to deaf man; and never, not ever you talk sex to eunuch. Him just get angry, sometime violent."*

This accounts for all that contumely and vituperation, the misrepresentation and deliberate neglect of real evidence; the rant, cant, and intolerant yap of the Defenders of the Archaeological Faith on the "Hall of Ma'at" website so heartily endorsed by Karen M. Romey in her contribution to the special section (characterized by "Ma'at" contributor Paul Heinrich as response "in a polite and understandable fashion").

Like academic dogs in the manger, you would deny others access to that which you are incapable of digesting yourselves. Spiritual attunement cannot be acknowledged. The ancients could not possibly have had knowledge or faculties you do not have. Everything must be kept locked up nice and safe in your little Darwinian box just in case someone pries open the lid and finds the Emperor's new clothes inside.

It's your loss—but, if you're actually interested in understanding why vast numbers of people refuse to accept your establishment expertise, this is why. Your special section will change nothing. It is an exercise in flawlessly sustained futility.

You also don't have your facts straight. A discipline exulting in minute detail (compare to Fagan) should be scrupulous in such matters. The Mars material you deride in *Mystery of the Sphinx* was never shown on TV. It was not part of the original NBC special, but was instead an addition to the expanded home video version. And I make it absolutely clear in that version that I am not endorsing either the Face or (especially) its putative relation to Egypt, but rather, consider the evidence supporting the notion provocative enough to merit inclusion. The Face, by the way, has NOT been "shown to be completely natural"; though NASA, the space equivalent of orthodox archaeologists, declares it so. There are a number of astronomers, geologists, physicists, and imaging experts, no less qualified than those at NASA, who do not accept that declaration. Since this is not our field, we happily leave that particular question open.

JOHN ANTHONY WEST

Dark Mission

THE SECRET HISTORY OF NASA

RICHARD C. HOAGLAND

My name is Richard C. Hoagland. I was a National Aeronautics and Space Administration (NASA) consultant to the Goddard Spaceflight Center in the post-*Apollo* era, and science advisor to Walter Cronkite and CBS News Special Events, advising CBS on the science of the NASA missions to the moon and Mars during the *Apollo* Program. I currently run the Enterprise Mission, an independent NASA watchdog and research group, which is attempting to figure out how much of what NASA has found in the solar system over the past fifty years has actually been silently filed out of sight as classified material, and therefore remains totally unknown to the American people.

My friend and colleague Mike Bara and I attempt the impossible in our book *Dark Mission: The Secret History of NASA*. We describe, and then carefully document, exactly what's been going on with NASA in terms of that classified data and information. It is not an easy task.

The predisposition of most Americans—even after the

Challenger and *Columbia* disasters and a host of other "missing" spacecraft—is to place NASA somewhere on par with Mother Teresa in terms of public confidence and credibility. This is, in major part, due to the average American's (to say nothing of the media's) inability to figure out a reason why NASA—ostensibly a purely scientific agency—would actually lie. NASA is, after all, holding high the beacon of our last true heroes, the astronauts.

However, even a hint that NASA—or, more precisely, its leadership—has been carrying out any kind of hidden agenda for over fifty years is met with disbelief at best. The vast majority of NASA's nearly 18,000 full-time employees are, in our analysis, innocent of the wrongdoing of the few that we are going to describe. The truth is that NASA was born in a lie, and has gone to extraordinary lengths to conceal the facts of its occult origins and its sensational discoveries on the moon and Mars.

Few people are aware that NASA was formed as a national defense agency adjunct empowered to keep information classified and secret from the public at large. Even fewer people are aware of the hard evidence that secret brotherhoods quietly dominate NASA, with policies far more aligned with ancient religious and occult mystery schools than the facade of rational science the government agency has successfully promoted to the world for almost fifty years.

Why was the Bush administration intent on returning to the moon as quickly as possible? What are the reasons for the current "space race" with China, Russia, and even India?

To even begin to fully appreciate what NASA has been quite consciously, deliberately, and methodically concealing from the American people and the world for all these years, you have to begin with NASA's turbulent past—specifically, an account of its origins in the increasingly dangerous geopolitical environment that Americans were thrust into in the wake of World War II.

The governmental institution known as NASA is a department of the executive branch, ultimately answerable solely to the president of the United States, an agency created through the National Aeronautics and Space Act of 1958. According to that act, NASA is, ostensibly, "a *civilian* agency exercising control over aeronautical and space activities sponsored by the United States." [Italics added.]

However, contrary to common public and media perception that NASA is an open, strictly civilian scientific institution is the legal fact that the space agency was quietly founded as a direct adjunct to the Department of Defense, tasked with specifically assisting the national security of the United States in the midst of a deepening Cold War with its major geopolitical adversary, the Soviet Union. It says so right in the original NASA Charter: "Sec. 305 . . . (i) *The [National Aeronautics and Space] Administration shall be considered a defense agency of the United States for the purpose of Chapter 17, Title 35 of the United States Code. . . .*" [Italics added.]

In another section of the act, this seldom-discussed *defense* responsibility—the ultimate undercutting of NASA's continuing public facade as a strictly civilian, scientific agency—is blatantly spelled out: "*Sec. 205 . . . (d) No [NASA] information which has been classified for reasons of national security shall be included in any report made under this section [of the act]. . . .*" [Italics added.]

Clearly, from this and the other security provisions incorporated in the act, what the Congress, the press, and the American taxpayers get to see of NASA's ultimate activities—including untouched images and data regarding what's *really* on the moon, on Mars or anywhere else across the solar system—is totally dependent on whether the president of the United States (and/or his legal surrogates in the Department of Defense and the "intelligence community") has already secretly classified that data. This is directly contrary to everything we've been led to believe regarding NASA for over fifty years now.

After NASA was formed, almost before the ink was dry on the bill that brought it into being (which, among many other detailed objectives, called for "the establishment of long-range studies of the potential benefits to be gained from, the opportunities for, and the problems involved in the use of aeronautical and space activities for peaceful and scientific purposes"), NASA commissioned a formal "futures study" into the projected effects on American society of its many planned activities (including covert ones).

Carried out as a formal NASA contract to the Brookings Institution—a well-known Washington, D.C.–based think tank— the 1959 study was officially titled "Proposed Studies on the Implications of Peaceful Space Activities for Human Affairs." The results of this multidisciplinary investigation were officially submitted to the administrator of NASA in late 1960 and (after the Kennedy administration was elected) to Congress in April 1961.

One area of unusual interest covered in the report—easily overlooked amid mountains of interminable statistics and analyses— was a quiet assessment of the near-certainty of a NASA discovery of intelligent extraterrestrial life: *"While face-to-face meetings with it [extra-terrestrial life] will not occur within the next twenty years (unless its technology is more advanced than ours, qualifying it to visit Earth), artifacts left at some point in time by these life-forms might possibly be discovered through our [NASA's] space activities on the moon, Mars, or Venus."* [Italics added.]

This quietly inserted subsection of Brookings is revealing on many levels, and it forms the documented basis of our case—that the NASA "you thought you knew" doesn't actually exist, and that NASA has been deliberately concealing and classifying its most significant discoveries because of "national security" rationales.

Brookings officially affirmed NASA's expectations that the agency would fly to nearby planets in the solar system, and would thus be physically capable, for the first time, of confronting "extra-terrestrials" *right in their backyard.*

Did any skeptics even know that this official document existed before we made it public in 1996? Beginning in the mid-1960s with unmanned spacecraft, NASA actually discovered its projected extraterrestrial artifacts—but the agency never got around to telling the rest of us!

NASA would clandestinely confirm with these earliest robotic probes, and then proceed to cover up, the first awesome remains of an extraordinary, solar-system-wide, ancient technological civilization on the moon—precisely as Brookings had predicted. Four years later, the *Apollo* Program would come to full fruition, and the lunar astronauts themselves would personally witness and extensively document, with tens of thousands of high quality photographs, from both lunar orbit and the surface, extraordinary "glasslike" structures on the moon! The *Apollo* crews would not only bring back rocks to NASA laboratories, but actual *samples* of the ancient technologies they found—for highly classified efforts at "back engineering."

At this point, a skeptic might well ask how we can be presenting valid, official NASA images of suppressed ruins and technology if the agency has spent so much time and energy over the past forty years covering them up. The answer is that after two generations, leaked images—displaying stunning details of ancient lunar structures arching overhead, as well as key alien artifacts that have been brought back—have also suddenly begun appearing on the Internet, on official NASA websites!

A small cadre of loyal NASA employees was witness to what actually went on, and agreed at that time to keep the secret in the interest of national security. Some of these NASA employees, apparently, have finally "seen the light"—that this continued deception, no matter what the legal rationale or national security implications, was fundamentally extra-Constitutional. Because of these true NASA heroes, real space history is about to officially begin, again.

Based on our analysis as presented in this book, it is also our opinion that NASA's entire lunar exploration program—culminating with the incredibly successful manned *Apollo* Project—was carefully conceived, from the beginning, as a kind of "alien reconnaissance" followed by an "alien artifacts retrieval" program.

Again, the intention to do just that was blatantly laid out in Brookings. We now believe that this is the reason President John F. Kennedy—reported to be "totally disinterested in space"—was quietly convinced to announce his historic decision to "send men to the moon . . . and return them safely to the Earth . . . within a decade" in May of 1961. This was widely believed, then and now, to be Kennedy's effort to demonstrate to the world the superiority of the American system, as opposed to Soviet communism. However, at the United Nations on September 20, 1963, the president suddenly issued a public invitation to the Soviets only two years into the *Apollo* "race" to the moon: an offer of a "cooperative, joint U.S./USSR lunar expedition."

Of course, if there were a "hidden agenda" to *Apollo,* this move would have revealed that the prime objective was not to beat the Soviet Union, but to covertly find and return samples of the incredibly advanced lunar technology that had been waiting on the moon for eons . . . and then to share them with the Soviets! Curiously enough, a mere two months following Kennedy's startling UN proposition, the president was killed.

The enthusiastic architects of the continuing NASA Brookings cover-up, in part, are the same heroes we have been encouraged to worship as some of the leading pioneers of our technological era. Their names are synonymous with America's achievements in space science and rocket engineering. In many cases, they are also men with secret pasts—Germans, Egyptians, Englishmen, and Americans, men at the very fringes of rational thought and conventional wisdom. These literal "fringe elements," then, are divided into three main groups inside the agency, as best as we can tell at present. For the purposes of this volume, we shall call them the "Magicians," the "Masons," and the "Nazis"—and deal with each group separately.

Each "sect" is led by prominent individuals, and supported by lesser-known players. Each has stamped their own agenda on our space program, in indelible but traceable ways. And each, remarkably, is dominated by a secret or "occult" doctrine that is far more closely aligned with "ancient religion and mysticism" than it is with the rational science and cool empiricism these men promote to the general public as NASA's overriding mantra.

Using commercially available celestial mechanics/astronomical software—programs like the popular "Red Shift" series (which uses the official JPL ephemeris as its database)—we have been able to establish a pattern of behavior on NASA's part that points to something as truly inexplicable as it is exotic: a bizarre internal obsession by the agency with three "gods" and "goddesses," reaching across the millennia directly from ancient Egypt—Isis, Osiris, and Horus.

It is these same three Egyptian gods (whose mythic story has been documented by many Egyptologists and authors) that are also key to understanding the history of the Masonic Order. It is this same mythology that is also at the heart of the belief systems of the NASA Magicians and Nazis as well. This ritual Egyptian symbolism, secretly practiced by NASA throughout these past five decades, publicly shows up only in its repeated, blatant choices of simple mission patch designs.

For instance, if one looks at the official patch for the *Apollo* Program, armed with our preceding "heads-up" regarding the bizarre NASA focus on all things "Egyptian," it becomes elemental to match the "A" (for "Apollo") as an actual stand-in for "Asar"—the Egyptian designation for "Osiris." This successful decoding of the hidden Egyptian meaning of the Apollo patch is redundantly confirmed—because "Asar/Osiris" is none other than the familiar Greek constellation of "Orion"—which is, of course, the background stellar constellation on the patch itself.

These continuing, inexplicable "secret society" manipulations inside NASA—not only of its personnel, but also of its major policies—have been going on since its Congressional formation, and all with this inexplicable "Egyptian focus." The evidence that NASA is "something other than the benevolent civilian science institution it pretends to be" is as overwhelming now as it is disturbing. In the years following JFK's assassination, when *Apollo* finally became an engineering reality, only nine successful *Apollo* missions to and from the moon were carried out; only six of those were actually lunar landings.

Then (apparently), a critical number and type of lunar artifacts were successfully identified, and returned to Earth by the *Apollo* crews—at which point the entire *Apollo* Program was abruptly terminated with *Apollo 17.* In our model, it was this successful completion of *Apollo*'s secret mission and agenda, and not Congressional budget cuts, that was the real reason for the abrupt cessation of America's historic journeys to the moon, and the primary reason no one has gone near the lunar surface for over thirty years.

All of which made the sudden announcement by President George W. Bush of a new White House/NASA program "to return to the moon by 2020," made at NASA headquarters on January 14, 2004, so incredibly intriguing. What did the Bush administration know, thirty years after the termination of *Apollo,* regarding what was waiting on the moon for human beings to return to? And was this why this administration mounted an *Apollo*-style program "on steroids"—as NASA administrator Mike

Griffin, specifically appointed by President Bush to head the new lunar return program—wryly termed it?

Was NASA's sudden interest in returning to the moon actually an effort to beat a host of other countries with the same goal? Countries that independently, suddenly announced their plans to go to the moon—countries like China, India, Japan, and Russia, and even the European Space Agency? Is this the start of a second space race, a race not for mere propaganda victories this time, but a much more important race, among a much wider field of players, for sole access to the scientific secrets the set of surviving lunar structures surveyed by *Apollo* must inevitably contain—which, to those who successfully decode what they discover this time, could mean the ultimate domination of the Earth?

It was just a few years after the start of the *Apollo* missions that even more extraordinary solar system ruins were first observed on Mars—beginning with images and other instrumented scans sent back by NASA's first Mars orbiter, the unmanned *Mariner 9* spacecraft, in late 1971.

This earliest robotic confirmation that there was also something "anomalous" on Mars paved the way for far more extensive observations when the first *Viking* orbiters and landers arrived half a decade later. The critical thing here is that, in direct contradiction to everything the press and the American people were being led to believe that NASA stood for—program transparency, open scientific inquiry, freedom of publication—the agency quietly and methodically covered up the most astonishing wonders it had found.

Many of NASA's consulting sociologists and anthropologists to Brookings (like Dr. Margaret Mead, whom I had the privilege of actually working with, in later years, at New York's Hayden Planetarium) had been warning NASA, even as Brookings was being researched and assembled, of "the enormous potential for social instability" if the existence of bona fide extraterrestrials—or even ruins they left behind—was officially revealed in the socially repressive and heavily religious environment of the late 1950s.

With those first Lunar Orbiter images taken of the moon, *everything*—the reality of ruins, their extraordinary scale, their obvious presence on more than one world in the solar system, how their builders vanished—was suddenly all too real. There *had* been a powerful, enormously encompassing, extraordinary solar-system-wide civilization that had simply disappeared, only to be rediscovered by NASA's primitive initial probes. A civilization that, it would turn out later, had been wiped out through a series of all-encompassing, solar-system-wide cataclysms.

The most disturbing part of Brookings to policy makers, however—even before these shattering discoveries were verified— was its thinly-veiled, authoritative warnings regarding what could happen to our civilization if NASA's 1950s-style "ET predictions" were confirmed:

"Anthropological files contain many examples of societies, sure of their place in the universe, which have disintegrated when they had to associate with previously unfamiliar societies espousing different ideas and different life ways; others that survived such an experience usually did so by paying the price of changes in values and attitudes and behavior. . . . The literal *disintegration of society*— simply from knowing that 'we're not alone.'" [Italics added.]

The Brookings discussion of the implications of such a crucial discovery also encompassed a critical second-level problem: What to do if the agency, at some point in the future, actually made such a momentous, world-changing confirmation of extraterrestrial intelligence next door? Or even of their surviving ruins and artifacts?

NASA's discussion of these problems before they occurred—and the draconian measures it was seriously considering—is revealing: *"Studies might help to provide programs for meeting and adjusting to the implications of such a discovery. Questions one might wish to answer by such studies would include: How might such information, under what circumstances, be presented to or withheld from the public, for what ends? What might be the role of the discovering scientists and other decision makers regarding release of the fact of discovery?"* [Italics added.]

Following the political tumult and excitement of the first successful *Apollo* lunar landings, the White House and NASA dramatically changed the direction of the entire space program—under the excuse of a lack of public interest and insufficient funding. The agency quickly dropped any pretense of following up on the *Apollo* Program with permanent bases on the moon, as well as indefinitely postponing all discussion and plans for going on to Mars.

Instead, under the now-proven lie of developing an economical, reliable, reusable space transportation system, and a "world-class" space research laboratory for it to resupply—the space shuttle and the International Space Station—NASA collaborated with the White House in a fateful set of decisions in the early 1970s that would consign American astronauts to endlessly circle the Earth for decades, while the moon—with stunning ruins and bits and pieces of a miraculous, preserved technology orbiting just a quarter of a million miles away—was totally ignored.

On February 15, 2001, FOX aired a widely advertised show titled *Conspiracy Theory: Did We Land on the Moon?* With this program, FOX removed the last weak link in NASA's ongoing, forty-year-old chain of overlapping cover-ups. It is our assertion that this "moon hoax" tale was carefully constructed as an elegant piece of professional disinformation—as a desperately required distraction from the *real* lunar conspiracy documented here, which was beginning to seriously unravel as early as 1996. I can personally testify that I was a firsthand witness to the true beginnings of "the moon hoax" far, far earlier than the 2001 FOX special—back in 1969, and in the heart of NASA itself!

The occasion was the unforgettable *Apollo* summer of Neil Armstrong and Buzz Aldrin's epic journey to the moon—the amazing July landing of *Apollo 11*. I, of course, had been deeply immersed in all aspects of our CBS coverage of the upcoming *Apollo 11* mission for months, as official science advisor to CBS News Special Events and chief correspondent Walter Cronkite.

For the actual flight of *Apollo 11,* I was assigned (at my own request) to the Downey, California, facility of the prime contractor for the *Apollo* Command/Service Module, North American Rockwell. I was there to personally oversee construction and special effects use of my pet project for our nonstop CBS coverage of "Lunar Landing Day"—a "walk-through solar system" constructed by North American Rockwell technicians specifically for myself and CBS in a huge, drafty aircraft hanger. It was in this miniature, re-created version of the solar system that I had successfully proposed that Walter Cronkite interview, via satellite, key engineers, project managers, and "special guests"—those who had built the *Apollo* spacecraft at North American or who had special knowledge in the realm of history and space—to comment on the historic legacy of the *Apollo 11* flight.

One luminary I was proud to bring before the cameras, to chat with Walter in New York regarding the extraordinary nature of events occurring that historic night, was Robert A. Heinlein, the dean of American science fiction. Decades earlier, Bob had cowritten the screenplay for *Destination Moon,* one of the first technically accurate film depictions of the lunar journey then unfolding on live television before a billion people all over planet Earth. As the successful author of a pioneering series of "juvenile" science fiction novels that, for the first time, introduced *realistic* space travel and engineering concepts to an entire generation of future NASA scientists and engineers, Heinlein had, almost single-handedly, "inspired the workforce" for the entire space program.

I must admit, I had a certain smug satisfaction that night, watching Bob Heinlein stroll through "the solar system," emphatically predicting to Walter and literally the world, via satellite, that "henceforth, this night—July 20, 1969—will be known as 'the Beginning of the True History of Mankind.'"

After the heady events of that unforgettable thirty-two hours— the landing; the eerie extravehicular activity (EVA), complete with ghostly television shots "live from the moon"; and then, after the crew had slept for a few hours for the first time on the moon, the successful liftoff of the *Eagle Lunar Module* and rendezvous with the *Command Module Columbia,* still in lunar orbit—CBS moved our unit up the street, to the Jet Propulsion Laboratory (JPL) in Pasadena. There we would cover the remainder of the flight, arriving at JPL right after the three *Apollo 11* astronauts blasted home toward Earth and their "splashdown" in the South Pacific three days later.

The reason was that NASA had another mission under way during the "Epic Journey of *Apollo 11*"—a flyby of two unmanned Mariner spacecraft past Mars, for only the second time in NASA's history.

With only one CBS Special Events Unit in California, to cover *all* of NASA's space activities on the West Coast in those years, it was up to our small group in Los Angeles—a producer, a correspondent, a couple of camera guys, maybe a couple of technicians, a makeup person, and me—to overlap our continuing coverage of *Apollo 11,* now originating from the Theodore von Kármán Auditorium at JPL, with new commentary covering the second unmanned NASA mission to do a flyby of Mars in history.

Mariner 6 was to cruise past Mars on July 31—recording television images, making spectral scans, conducting remote atmospheric measurements—just ten days after the *Columbia* left lunar orbit, heading for the Pacific Ocean. Our arrival at JPL on the afternoon of July 22, in preparation for this *Mariner 6* flyby, was heady stuff for a twenty-three-year-old network science consultant, as this was my first "in-person" tour to cover an actual live mission.

The circumstances of my first flyby live from JPL are etched indelibly in my brain, if for no other reason than it was the moment when television lightning struck. One morning our executive producer, Bob Wussler, suddenly decided to put *me* on the air across the entire CBS television network to explain the upcoming Mariner flyby to the nation! How could one ever forget their first professional network television appearance—and their first official network commentary for a NASA mission flying by Mars, no less? But for the life of me, I can't remember a thing I said that morning. I do remember that I literally had to borrow a sport coat and tie from one of the floor crew for my first appearance on network television.

And, I vividly remember a bizarre scene that happened only a couple of days before at JPL, as we arrived.

It was controlled bedlam. Close to a thousand print reporters, television correspondents, technicians, special VIPs, as well as half the staff at JPL itself, were all attempting to register for the limited seating in the (relatively) small von Kármán Auditorium, which had been the scene for all live network coverage of JPL's previous extravaganzas ever, since Explorer 1 had been placed in orbit by a U.S. Army/JPL team one January night in 1958.

This warm July afternoon only eleven years later, it seemed that everyone was in a mad scramble—simultaneously—to register at the lobby desks specifically set up for members of the press, trying to grab the limited number of press kits on the mission, and then nail down a seat in the auditorium beyond.

It was at this point, as I was drifting around von Kármán, trying to spot where the CBS anchor desk was positioned, that I noticed something strange. Even to my untrained eye, it looked out of place:

a man, wearing jeans and a long, light-colored raincoat (it was typical Los Angeles weather outside—so, why the coat?). This man, wearing one of those floppy "great coats" that cowpunchers used to wear in old Westerns, complete with a dark leather bag slung over one shoulder, was slowly, methodically, placing "something" on each chair in von Kármán.

As he got closer, I suddenly realized that he was accompanied by a more conventionally dressed representative from JPL itself—coatless, in white shirt and black tie—the second figure was, in fact, none other than the head of the JPL press office, Frank Bristow. In the midst of all the commotion, why was Bristow—again, the head of the JPL press office—*personally* squiring this very out-of-place individual around the auditorium?

Then, as if that wasn't mystery enough, Bristow began moving "great coat guy" back out to the cramped pressroom area beyond the glassed-in foyer of the auditorium. There, in an office where space correspondents like Walter Sullivan (*New York Times*), Frank Pearlman (*San Francisco Chronicle*), Jules Bergmann (ABC), and Bill Stout (our local guy from CBS) hung out, and wrote their leads and copy after each formal press briefing held in von Kármán itself, a handful of reporters were now being introduced, again by Bristow, to "great coat guy." Why was the official head of the JPL press office doing this?

I soon had my answer.

As Bristow watched approvingly, his "guest" proceeded to hand each available reporter a copy of whatever he'd been putting on the seats back in the auditorium.

As I opened up the handout, something yellow and silvery fell on the tile floor. It was a shiny American flag, maybe four inches lengthwise, made of aluminized mylar. I turned to the couple of mimeographed pages and began to read—and couldn't believe my eyes.

The date was July 22, 1969. The three *Apollo 11* astronauts, Neil Armstrong, Buzz Aldrin, and Mike Collins—two of whom had just successfully walked on "the frigging moon" and wouldn't splash down in the South Pacific for two more days—were still halfway between Earth and the "Sea of Tranquillity." Yet here, someone with an obvious "in" to JPL was handing out a mimeographed broadsheet to all the *real* reporters . . . claiming "NASA has just faked the entire *Apollo 11* Lunar Landing . . . on a soundstage in Nevada!"

And, if that wasn't weird enough, this individual was being *personally* escorted around von Kármán by none other than the head of the JPL press office himself! I did what I saw the other veterans do: I casually threw the two pages in the trash and tucked the shiny flag into my notebook . . . but the seed had been planted.

Looking back, based on all our hard-won knowledge of what is really "out there" in the solar system, and experiencing the outrageous lengths NASA will go to keep "the secret," I can now put the pieces together.

This was an official Op—Bristow's job was to make sure that all the national reporters covering NASA at least saw what was handed out that afternoon, complete with shiny flag to act as a mnemonic device to trigger the memory of what was in the pamphlet long after it was history. Sooner or later, a percentage of those who read it that afternoon at JPL would write it up—as a quirky angle on the far-too-dry official tale of *Apollo 11*.

In this way, it would become a naturally reproducing meme— "a unit of cultural information, such as a cultural practice or idea, that is transmitted verbally or by repeated action from one mind to another"—which is exactly what NASA apparently intended to plant at JPL that afternoon, deliberately to "infect" American culture with the story that "the moon landing was all a fake!"

Was this all some far seeing "back-up plan" if, in some point in the future, it started to emerge why the astronauts had *really* gone to the moon?

FOX, the "fair and balanced" network, activated the meme in 2001 with the *Did We Land on the Moon?* special. There, waiting in the wings, was a neatly packaged thirty-year-old "conspiracy theory" perfectly gift-wrapped for those finally beginning to "disbelieve" in NASA. An officially concocted "inoculation" against troublemakers who would one day place before many of those same national reporters a set of embarrassing official *Apollo* photographs, asking the crucial question: "What did NASA *really* find during its *Apollo* missions to the moon?"

SEVEN

History and Celestial Time

DOES PRECESSION CAUSE THE RISE AND FALL OF CIVILIZATION?

WALTER CRUTTENDEN

Discoveries like the ancient Greek Antikythera Device (1,500 years before the invention of precision geared devices), the Baghdad Batteries (two thousand years before Volta "invented" the battery), or dental and brain surgery artifacts found in ancient Pakistan (eight thousand years out of historical sequence) appear "anomalous" within our current paradigm of history. However, they are not unexpected according to the ancient cyclical view.

Giorgio de Santillana, a former professor of the history of science at MIT, tells us that most ancient cultures believed consciousness and history were not linear but cyclical, meaning they would rise and fall over long periods of time. In his landmark work *Hamlet's Mill,* Giorgio de Santillana and coauthor Hertha von

Dechend showed that the myth and folklore of over thirty ancient cultures around the world spoke of a vast cycle of time with alternating Dark and Golden Ages that move with the precession of the equinox. Plato called this the Great Year.

Although the idea of a great cycle timed by the slow precession of the equinox was common to multiple cultures before the Christian era, most of us were taught that this is just a fairy tale— there was no Golden Age. However, an increasing body of new astronomical and archaeological evidence suggests that the cycle may have a basis in fact. More important, understanding the cycle might provide insight into where society is headed at this time and why consciousness may be expanding at an exponential rate. Understanding the cause of precession is key to understanding the cycle.

The standard theory of precession says it is principally the moon's gravity acting upon the oblate Earth that must be the cause of the Earth's changing orientation to inertial space, otherwise known as the "precession of the equinox." However, ancient sources say that the observable of an equinox slowly moving or "precessing" through the twelve constellations of the zodiac is simply due to the motion of the solar system through space (changing our viewpoint from Earth). At the Binary Research Institute, we have modeled a moving solar system and found that it does indeed better produce the precession observable (the sun's motion relative to the fixed stars as seen from Earth), and resolves a number of solar system anomalies such as the uneven distribution of angular momentum within the solar system and the variable rate of precession.

Beyond the technical considerations, a moving solar system might provide a logical reason why we have a Great Year with alternating Dark and Golden Ages. That is, if the solar system carrying the Earth actually moves in a huge orbit, subjecting the Earth to the electromagnetic (EM) spectrum of another star or

EM source along the way, we could expect that this would affect our magnetosphere, ionosphere, and—indirectly—all life, in a pattern commensurate with that orbit. Just as the Earth's smaller diurnal and annual motions produce the cycles of day and night and the seasons (both due to the Earth's changing position in relation to the electromagnetic spectrum of the sun), so might the larger celestial motion be expected to produce a cycle that affects life and consciousness on a grand scale.

The hypothesis for how consciousness would be affected in such a celestial cycle builds on the work of Dr. Valerie Hunt, former professor of physiology at UCLA. In a number of studies, she has found that changes in the ambient electromagnetic field (which surrounds us all the time) can dramatically affect human cognition and performance. In short, consciousness is affected by immersion in electromagnetic fields. Consequently, the concept behind the Great Year or cyclical model of history, consistent with myth and folklore, is based on the sun's motion through space, subjecting the Earth to waxing and waning stellar fields (all stars are huge generators of electromagnetic spectrum), resulting in the legendary rise and fall of the ages over great epochs of time.

We looked at some of the ancient myths about rising and falling ages tied to the precession cycle, explored current precession anomalies, outlined a dynamic solar system model that better explains the precession observable, and suggested a hypothesis for how a change in proximity to stellar-generated electromagnetic fields might be the mechanism that induces cyclical changes on Earth. We would here like to use this model as a guide to better understand where we have been in terms of consciousness and ancient civilizations in the past, and more important, where we are going in the future. As Graham Hancock stated, this "new—or very old—approach to the greatest problems of human history" could be a "key to the mystery of the ages."

Historical Perspective

Current theories of history generally ignore myth and folklore and do not consider any macro-external influences on consciousness. For the most part, modern historical theory teaches us that consciousness or history moves in a linear pattern from primitive to modern with few exceptions. Some of its tenets include:

- Mankind evolved out of Africa.
- People were hunter-gatherers until about five thousand years ago.
- Tribes first banded together for protection from other warring parties.
- Written communication must precede any large engineered structures or populous civilizations.

The problem with this widely accepted paradigm is that it is not consistent with the evolving interpretation of recently discovered ancient cultures and anomalous artifacts. In the last hundred years, major discoveries have been made in Mesopotamia, the Indus Valley, the Asian plains, South American sites, and in many other regions that break the rules of historical theory and push back the time of advanced human development. Specifically, they show that ancient man was far more proficient and civilized nearly five thousand years ago than he was during the more recent Dark Ages of just a thousand years ago.

In Caral, an ancient complex on the west coast of Peru, we find six pyramids that are carbon-dated to be 4,700 years old, a date contemporaneous with the Egyptian pyramids and rivaling the time of the first major structures found in the so-called cradle of civilization in Mesopotamia. However, Caral is an ocean away from the "cradle," and we find no evidence of any writing or weaponry, two of the so-called necessities of civilization. At the same time, we do find beautiful musical instruments, astronomically aligned structures, and evidence of commerce with distant lands. Clearly, such sites defy the standard historical paradigm—but what is stranger still is that so many of these civilizations seemed to decline en masse.

In ancient Mesopotamia; Pakistan; Jiroft, Iran; and adjacent lands, we see knowledge of astronomy, geometry, advanced building techniques, sophisticated plumbing and water systems, incredible art, dyes and fabrics, surgery, medicine, and many other refinements of a civilized culture that seemed to arise from nowhere, yet were completely lost over the next few thousand years. By the time of the worldwide Dark Ages, every one of these civilizations had turned to nomadic ways of life, or to dust. Near the depths of the downturn, there were ruins and little else to be found. And in some areas where larger populations still remained, such as throughout parts of Europe, poverty and disease were often rampant, and the ability to read, write, or duplicate any of the earlier engineering or scientific feats had essentially disappeared. What happened?

While records of this period are still very spotty, the archaeological evidence indicates consciousness, reflected as human ingenuity and capability, was greatly diminished. We just seemed to have lost the ability to do the things we used to do. Ironically, this is just what many ancient cultures predicted. The world's foremost Assyriologist, Stefan Maul, shed light on this phenomenon in his Stanford Presidential Lecture when he told us that the Akkadians knew they lived in a declining era; they revered the past and tried to hang on to it, but at the same time lamented and predicted the Dark Ages that would follow. His etymological studies of cuneiform tablets show that the ancient words for "past" have now become our words for "future," whereas their words for the "future" have now become our words for the "past." It is almost as if mankind orients his motion through time depending on whether he is in an ascending or descending age.

We find this principle of waxing and waning periods of time depicted in numerous bas-reliefs found in ancient Mithraic temples. The famed Tauroctany (or bull-slaying) scene is often surrounded by two boys, Cautes and Cautopetes. One holds a torch up on one side of the zodiac, indicating it is a time of light, the other holds a torch down on the other side of the zodiac, indicating it is a time of darkness. As the accompanying chart will show, these time periods correspond with the Vedic description of when the Earth goes through periods of rising and falling consciousness.

Jared Diamond, the well-known historian anthropologist and author of *Guns, Germs, and Steel,* makes a good case that it is primarily local geographic and environmental advantages on the planet Earth that determine which group of humans succeeds or fails versus another. Those that have the steel, guns, and bad germs win. While this helps explain many regional differences of the last few thousand years, it does not address the macro trends that seemed to have affected all cultures (including China and the Americas) as they slipped into the last worldwide Dark Age.

The cyclical or Great Year model overlays and augments Jared Diamond's observations, giving a reason for the widespread downturn. It suggests that it is not just the geography and environment of man on Earth that determines his relative success but it is also the geography and environment of the Earth in space that affects mankind on a vast scale. Just as small celestial motions affect life over the short-term, so do large celestial motions affect us over the long-term.

Understanding that consciousness may indeed rise and fall with the motions of the heavens gives meaning to ancient myth and folklore and puts anomalous artifacts such as the Antikythera

Device into an historical context that makes sense. It speaks to why so many ancient cultures might have been fascinated with the stars, and it provides us with a workable paradigm in which to understand history. It could also help us identify the forces that propelled the Renaissance, and that may be accelerating consciousness in the current era. Myth and folklore, the scientific language of yore, provide a colorful look at consciousness throughout the different ages.

Character of the Ages

The Greek historian Hesiod tells us of the wonderful nature of the last Golden Age, when "peace and plenty" abounded. Hopi myths tell us of cities on the bottom of the sea. Typically ancient peoples broke the great cycle into an ascending and descending phase, each with four periods. For example, the Vedic or Hindu culture tells us that when the autumnal equinox moves from Virgo to Aires, we go through the ascending Kali, Dwapara, Treta, and Satya (Golden Age) Yugas before slowly declining in reverse order as the equinox completes its journey. The Greeks and other early Mediterranean civilizations used similar periods and labeled them the Iron, Bronze, Silver, and Golden Ages. More distant cultures, such as the Maya or Hopi, used still other names, such as "worlds" or "suns," and labeled them "fourth or fifth" to identify the recent epochs.

A relatively modern proponent of the cyclical system was the Sanskrit sage Swami Sri Yukteswar, author of *The Holy Science*. He taught that the position of our solar system relative to another star now indicates that we are in recent transition from the lowest material age, the Kali Yuga, into the electrical or atomic age, the Dwapara Yuga. In this period, it is said that we begin to see the world as more transparent as we move from an awareness of self as a physical body in a strictly physical universe, to an awareness that we are something more, living in a universe filled with subtle forces and energies.

The technological discoveries of the laws of gravity, electricity, and magnetism, just in the last few hundred years, give this idea credence—and the trend is accelerating. In the last century it has even been discovered that physical matter is not really solid at all. We have found that it is made of molecules, and that these are in turn made of atoms, which are themselves constituted of 99.9 percent empty space. The little bit of matter that does exist in the heart of the proton and neutron is now thought to be principally vibrating energy, at least according to the latest version of string theory. Indeed, reality is looking more and more ethereal, just as the hoary Vedas predicted.

Ages beyond the present are difficult to grasp, because a lesser consciousness cannot behold a greater consciousness any more than a cup of water cannot hold the ocean—so we tend to extrapolate the past material view of things when envisioning the future. Yet the Oriental teachings about cycles indicate that this is just a passing phase. They say the real trend is toward a godlike state where the physical is but a manifestation of something from the other world. And so it seems when we read Greek mythology or pages of Vedic scripture.

The Silver Age or Treta Yuga, the third age (from the bottom) is the Greek "age of the demigods," or, to the yogis of India, the age of divine magnetism and the mind. While this is a difficult concept to grasp, consider the story of Babel.

Supposedly, before Babel (pre–3100 BCE in the last descending Treta Yuga) humanity spoke with one tongue and communed freely with nature. The Hebrew scriptures tell us that mankind began to build "towers," and then languages were "confused" and people could no longer understand one another (Genesis 11:1–9). In the standard theory of history this story makes no sense, but in the cyclical model it has great meaning. It would have occurred around the time of the first tower buildings in ancient Mesopotamia, probably between 3500 BCE and 3000 BCE.

This is precisely around the time (3100 BCE) when, according to Sri Yukteswar, the world declined from the descending Treta Yuga into the descending Dwapara Yuga, a time when clairvoyance and telepathy were lost. We learn from Paramahansa Yogananda, another proponent of the yuga cycle and the famed author of *Autobiography of a Yogi,* that this time will come again in the year 4100 CE, when we pass from the ascending Dwapara and into the ascending Treta Yuga. He tells us that at this time there will once again be a "common knowledge of telepathy and clairvoyance." Perhaps then we will better understand the meaning of the ancient myths.

The Treta Yuga is said to be the age of levitation, telepathy, a time of the shamans and wizards of old, when tremendous psychic and mental abilities were common, truly an "age of the demigods." We've all heard stories about the mythical powers of the saints and sages who had these gifts. Now seen as rare, the majority of people don't take these reports seriously or realize that we, too, might have this same latent ability in a higher state of consciousness. Yet, this is exactly what the ancients told us. In fact, Christ was quoting the far more ancient Hebrew scriptures when, in the depths of the last Dark Age, he said, "Is it not written in your law, I said, Ye are gods?" He himself embodied this consciousness when he said, "He that believeth on me, the works that I do shall he do also."

The final stage in the cycle of time is the Golden Age or Satya Yuga. It is considered the highest time on Earth. If the Treta or Silver Age is inconceivable to us today, then the Golden Age must sound like a myth or a dream. The Greeks called it the "age of the gods," and the myth and folklore of the Vedas and ancient Egypt hint that this was a time when gods literally walked the Earth, and most of mankind lived in perfect harmony with nature and the heavens. While there now remains very little physical evidence of this long-ago period, we do find that virtually every ancient megalithic construction prior to the year 1500 BCE seems to be oriented toward some astronomical or cardinal point. Going back further, there are signs that multiple structures may have been aligned to mirror constellations or the larger heavens.

The Golden Age is said to be a time when we could perceive and communicate with astral or causal realms, and directly know God without the intermediacy of any religion. Again, this sounds like little more than a fairy tale given our current state of consciousness, but it is a theme common to ancient peoples who spoke and wrote of the long-lost higher ages.

Predictive Value

Admittedly, the higher ages sound incredible, but we hope to show evidence at the next Conference on Precession and Ancient Knowledge (CPAK), and through future papers, books, and films, that the cycle has a basis in fact, driven by the solar system's motion through space. Just as the seasons of the year, caused by the Earth's orbit around the sun, can be forecast in time (through calendars and various astronomical means), so can the seasons of the Great Year be calculated by the slow precession of the equinox.

The cyclical model is not only precisely measurable (by monitoring the annual change in the precession rate, now about 50.29 arc seconds per year) but I believe that it has predictive value. There are many changes we can expect over the next few decades to few thousand years as we progress through the Great Year. Research into these changes is based on cross-interpretations of myth and folklore, extrapolation of trends, and interviews with futurists.

During the current transition from the Kali Yuga (of gross material consciousness) to the early Dwapara Yuga (where an awareness of energy and finer forces will be dominant), we are manifesting our heightened awareness and increasing ingenuity through an endless array of technology that allows us to annihilate the barriers of time and space. We can now fly just about anywhere on the globe within the time it takes the planet to make one spin on its axis. Likewise, we can instantly communicate with someone on the other side of the Earth and send them a picture or video of almost any event, real time. All these things were not only impossible but also unthinkable just a hundred years ago.

Underlying this trend, there is actually a greater concern for nature. We will see, more and more, a return to living in tune with Mother Earth—and it will be facilitated by greater understanding and thinner technology. As technology becomes something hidden in the background, we can expect some amazing changes. For example, while we currently still need antennas to transmit communications (and soon power) or silicon to compute or store information, even these may be outmoded in the future.

Physicist John Dering (a CPAK regular) has speculated that given the trend of computer power, sometime in the not too distant future we will develop interface devices that allow us to pick up the waveforms captured by trees or the antennae of bugs, and we may be able to tap into and decipher all the information (waveforms) that have ever passed by a rock or any inanimate object in the landscape. Could it be that our ancient ancestors better understood the subtle qualities of stone?

Another CPAK author, John Burke, has already shown that ancient cultures had a tremendous knowledge of electromagnetism as evidenced by the outer stones at Avebury, where he has demonstrated that all of the standing stones' magnetic poles are identically aligned. He has also shown that some Indian shamans in the

American West can find areas of high electrostatic charge or geo-physical discontinuities just by feel. They use these areas for healing purposes. Contemplating these ideas gives new meaning to the stories of our ancient ancestors. Understanding their wisdom may be the key to understanding our future.

EIGHT

The Orion Key

UNLOCKING THE MYSTERY OF GIZA

SCOTT CREIGHTON

Upsetting the Apple Cart

It has long been the view of mainstream Egyptologists that the siting of each of the main pyramid structures at Giza was determined purely upon the wishes of the pharaoh and the practicalities and logistics of a particular site. It is further held that the King's decision for the siting of his pyramid gave little or no consideration to structures that had gone before or that would come after. Each pyramid at Giza—so the conventional view asserts—was effectively constructed as a discreet royal funerary complex by each successive pharaoh, and was done so without reference to any long-term master site plan. In short, we are told that the pyramids at Giza were constructed as "singularities" and that there existed no grand architectural scheme; no grand plan.

This conventional view of the pyramids at Giza runs contrary to the work of Robert Bauval who, in 1994—in partnership with

Adrian Gilbert—published *The Orion Mystery,* his first book, which presented the radical hypothesis that the pyramids at Giza were constructed as symbolic representations of the three Belt stars of the constellation of Orion. By advocating such a hypothesis, Bauval was invoking the almost heretical notion that each of the pyramids at Giza were constructed as component parts of a long-term project; a multigenerational master plan that involved the Belt stars of the constellation of Orion.

It is unsurprising that in making such a bold hypothesis, Bauval quickly incurred the wrath of the academic establishment. With a few notable exceptions, Egyptologists remained largely unconvinced of Bauval's proposal, dismissing much of the cultural evidence he presented from the ancient Egyptian writings known as the Pyramid Texts, which provided considerable textual support to his work. The Egyptologists demanded that Bauval present conclusive evidence in support of his Orion hypothesis before they would even remotely consider overturning more than a century of Egyptological consensus that staunchly regarded the Giza pyramids as three discreet royal funerary complexes built entirely independently of each other.

Robert Bauval, however, remains steadfast in his view. Almost fifteen years after publishing *The Orion Mystery,* I met with Bauval in the shadow of the Great Pyramid and asked him his opinion concerning a very obvious anomaly at Giza. "If Khufu was the first pharaoh of the Fourth Dynasty to build at Giza," I put it to him, "why, then, didn't Khufu construct his Great Pyramid on the prestigious, high ground in the center of the plateau?"

Bauval turned to me with his characteristic enigmatic smile, and replied, "Because, Scott—there was a plan."

In *Keeper of Genesis/Message of the Sphinx,* the authors show how this "grand plan" may also have included the Great Sphinx at Giza, which they hypothesize might have been designed as a reflection of the lion constellation of Leo in the eastern sky circa 10,500 BCE. It seems that the designers of the Great Sphinx may have used the constellation of Leo as the underlying template for the design of this most recondite of all Giza structures. On this basis, might it not then be possible that the designers may have used the Orion's Belt stars not simply to lay the pyramids on the ground in a near identical pattern to the Belt stars, but to *also* use the Belt stars as the underlying template from which to generate the *actual dimensions*—the base shape—of all three main Giza pyramids?

It hardly needs to be stated, but if it can be demonstrated that the Orion's Belt stars can indeed be used to simply and easily generate three bases whose dimensions proportionally match those of all three main Giza pyramids, then this would provide strong circumstantial evidence in support of Bauval's Orion hypothesis.

Yet can this be done? Is it possible that the three Orion's Belt stars can—when applying a very simple geometrical technique— be used to produce three bases that perfectly match the bases of the three Great Pyramids at Giza? We will shortly investigate this idea, but for the moment let us consider some further "anomalies" that exist at Giza, for which there exist no convincing conventional answers.

Anomalies Abound

As already mentioned, if the final resting place of the pharaoh was simply a matter of his own personal choice, then we have to ask ourselves why Khufu should have chosen to locate his Great Pyramid at the very edge of the northeast corner of the Giza Plateau. On the surface, this would not seem a particularly odd choice. Looking closer, however, we find that the high, dominant ground lies in the *center* of the plateau where Khafre's Pyramid stands. Not only is this central area the commanding ground on the plateau, but it also benefits from having a "natural causeway" that runs from the area of the Sphinx up to the east face of Khafre's Pyramid. Even today, Egyptologists refer to this area as the "gateway to Giza."

Why then, we ask, would Khufu choose to construct a quite monumental *artificial causeway* deep into the valley, when a natural causeway *already existed* up to the high, commanding ground of the plateau and which—had Khufu utilized it for himself—would have considerably eased the construction burden of his funerary complex?

Furthermore, by failing to claim for himself the highest ground on the plateau, Khufu would have been fully aware that he was leaving the door open for some future pharaoh to trump his own achievement by potentially building a higher pyramid in that area, which, as matters transpired, is precisely what happened—Khafre's Pyramid appears larger than the Great Pyramid due to it having been built on the central high ground of the plateau.

The ancient Egyptians were—first and foremost—very practical builders. For Khufu to have decided against building his pyramid in the most *practical and prestigious* location on the plateau defies logic, and is entirely inconsistent and anomalous. Why would Khufu risk having the magnificence of his own grand structure eclipsed by leaving the Giza door wide open for a future pharaoh to come along and surpass his own achievement? It makes no sense.

And why did Khafre—the builder of the second Giza pyramid—decide to return to Giza when his predecessor, Djedefre, did not build at Giza like his father, Khufu, but instead opted to build his pyramid at Abu Rawash? Indeed, Khafre was the first king of the Fourth Dynasty to "co-locate" at the site of another pharaoh (Khufu). Why did Khafre (and indeed, Menkaure) decide to build at Giza where—from a religious perspective—it would have seemed more consistent to have co-located with Djedefre (the first king of the emerging solar cult to have the name of the sun god Re incorporated into the royal cartouche) at Abu Rawash?

And finally, we have Menkaure's Pyramid—the smallest of the three Great Pyramids at Giza. Once again it defies basic human instinct that Menkaure would have wished to locate his own infinitely smaller structure in the shadow of the Great Pyramid and Khafre's Pyramid, where its relatively diminutive stature would only have been exacerbated by its close proximity to its two illustrious neighbors. Had Menkaure sited his own structure at a virgin site, he could have avoided the ignoble fate that assuredly awaited his pyramid by deciding to build at Giza. Why would Menkaure commit to Giza in the full knowledge that the pyramid he planned to build there would forever suffer the ignominy of having failed to meet the high standards set by his predecessors? Again, it makes little sense.

Another very obvious anomaly pertains to the two sets of three smaller satellite pyramids known as the "Queen's Pyramids." Three of these structures run in a north/south line to the east of Khufu's Pyramid, whilst the other set of three run in an east/west line just south of Menkaure's Pyramid.

Khafre's Pyramid has no queen's pyramids at all. Why would this be when it is known that Khafre actually had *five* wives? Five queens, and yet not a single queen's pyramid is located at Khafre's Pyramid! There seems no consistency, no logic to this anomaly.

Of course, all of these anomalies, contradictions, and motives are simply and easily explained by the prior existence of a grand, overall plan; a *strict* plan that the ancient Egyptians of the Fourth Dynasty—beginning with Khufu—felt compelled, indeed "duty-bound," to implement on the ground at Giza.

From a purely logistical point of view, it would make sense to build Khufu's Pyramid first, since to commence the plan by building on the central high ground of the plateau would have presented a considerable obstruction to later construction. Building *had* to begin in the northeast of the plateau to accommodate the logistics of building the structures that would come later in the plan. Given this constraint, Khufu—being the first to build at Giza—had little choice but to build his pyramid on the lower ground in the northeast of the plateau. However, had there been no master plan for later structures, Khufu would undoubtedly have taken full advantage of the prestigious high ground for his own pyramid. Khufu's peculiar actions tell us that his "hands were tied"—he was constrained by the demands and logistics of a *greater plan;* a plan he was seemingly powerless to influence.

In Search of the Master Key

Whether the pyramids at Giza were constructed as "singularities" or as part of some greater scheme, it is not unreasonable to surmise that the builders would have followed plans of some kind. Alas, no plans of any kind have ever been recovered.

However, even in the absence of any actual plans, it may still be possible to uncover the blueprint the ancient builders may have used to lay out and define the dimensions of the Giza monuments—and we can perhaps achieve this by reverse-engineering the geometry of the in situ structures. The logic is simple—if the Giza pyramids were constructed as part of some overall "master plan," then it naturally follows that there would exist some guiding principle common to all three pyramids. Finding this "key" would hopefully lead to the underlying design imperative—the blueprint—and show that the pyramids were not simply placed in random fashion on the whim of three successive pharaohs.

Over the years, the Internet has spawned a veritable plethora of geometrical-mathematical offerings that attempt to seek the answer to the question of how the three main Giza pyramid base dimensions were arrived at and how the other satellite pyramids were placed. So far, however, none of these offerings provides a complete, cohesive solution to this particular question. The master "key" to the Giza blueprint remains frustratingly elusive.

It was stated in the opening to this article that Robert Bauval has long argued that the three main Giza pyramids were built and laid out on the ground at Giza as symbolic representations of the three Belt stars of the Orion constellation—not a perfect center-to-center correlation, but very close. Assuming, then, that Bauval's hypothesis is correct, could it then be possible that the Orion's Belt stars might reveal to us other aspects of the Giza pyramid design? Might it be possible that the Belt stars could have been used to define the actual geometrical base dimensions of the three main Giza pyramids? In short—could the Orion's Belt stars be the elusive design "key"; the underlying Giza design imperative sought by so many for so long?

Well, let's see.

The Orion "Geo-Stellar Fingerprint"

Any three non-linear points (in this case the three stars of Orion's Belt) can be used to generate three square bases, simply and easily, using a technique that has come to be known as "geo-stellar finger-printing." Orion's Belt stars can be used to produce their very own "geo-stellar fingerprint."

The dimensions of the three main pyramids at Giza are shown to proportionally agree with the three bases produced by the Orion Geo-Stellar Fingerprint—a clear connection between the Orion's Belt stars and the pyramids at Giza. Furthermore, what we also find is that three lines can be passed through each of the three main pyramids of the Geo-Stellar Fingerprint that allow us to also place the Queen's Pyramids of Menkaure and Khufu.

It seems somewhat ironic that the mathematical "solution" to the Giza layout sought by so many over so many years was sought by some individuals to *disprove* Bauval's Orion hypothesis, and yet what we now find is that the Orion's Belt stars actually held the key to the mathematical design solution—the *Giza blueprint*—that these individuals had been so keen to discover.

Yet the Orion influence at Giza does not end there. There is more—much more.

Whilst the Orion Geo-Stellar Fingerprint presents highly compelling evidence on its own of a direct (mathematical) connection between the three Giza pyramids and the Orion's Belt stars, independent corroboration of such an association is desirable to further strengthen the hypothesis. If it could be shown that *other* pyramid structures at Giza *also* have a demonstrable connection with the Orion's Belt stars (with the obvious placement of these structures within the Orion Geo-Stellar Fingerprint) then this would provide the independent corroboration required to prove the Orion-Giza hypothesis—if not conclusively, then certainly beyond any reasonable doubt.

Queens of Precession

It would seem that such independent corroboration can be found in the relative placement of the two sets of three "Queen's Pyramids" on the plateau.

Before presenting this corroborating evidence of an Orion association with the two sets of Queen's Pyramids, it is important to first take a moment here to understand the astronomical phenomenon referred to by astronomers as "precession."

Most of us are familiar with the Earth's daily (diurnal) rotation of twenty-four hours. We are also familiar with its second motion, its annual 365-day rotation (orbit) around the sun. The Earth, however, possesses a *third,* much less perceptible motion known as "precession."

When we look at the night sky, we observe the stars slowly rotating in a continual east to west direction. Over some 13,000 years, however, the stars in our night sky actually slowly drift in a west to east (retrograde) direction before stopping and then, over a further 13,000 years, drift slowly back to their point of origin—like the swing of a clock's pendulum. The conventional view of this "precessional drift" is that it occurs as a result of a very slow wobble of the Earth as it rotates around its axis.

The end result of this "precessional drift" is to slowly shift the rising and setting points of stars on the horizon over a long period of time—approximately one degree of precessional drift occurs every seventy-two years. One beneficial aspect of this precessional drift is that it can be utilized as a mechanism for "recording" time, for marking important dates. For example, by aligning two stone obelisks with a setting star precisely due south (180 degrees), we are effectively marking that specific moment (date) in time. Over the passing years and decades, the target star continues its slow drift along the horizon, leaving the alignment with the obelisks far behind. The obelisks no longer align with the target star on the horizon but now serve as a "marker"—an astronomical record—of when the alignment was made with the target star.

By then observing where the target star sets (or rises) today—for example, at 192 degrees—we can then extrapolate that it has drifted some 10 degrees from its original obelisk alignment of 180 degrees which, in turn, tells us that the alignment was created approximately 720 years in the past, since seventy-two years equals approximately one degree of precessional drift.

Similarly, when we now consider the pyramid structures at Giza, we find a very clear and unequivocal alignment of those structures with the Orion's Belt stars as they appeared on the southwest *horizon* circa 10,500 BCE. Specifically, this alignment involves the smallest of the three Great Pyramids—the Pyramid of Menkaure—and its stellar counterpart, the belt star Mintaka in Orion's Belt. Mintaka is the "target" star through which the Pyramid of Menkaure (our "obelisk") is aligned.

Circa 10,500 BCE, the star Mintaka set close to the southwest horizon 212 degrees azimuth (212 degrees clockwise from due north). When considering the Giza pyramids, we find that the alignment from the apex of Khafre's Pyramid through the apex of Menkaure's Pyramid is also 212 degrees azimuth.

At the very same moment, the three Queen's Pyramids of Menkaure are placed in a horizontal line close to the southwest horizon, thereby mimicking the arrangement of the three Belt stars which are similarly arranged at that precise moment before setting on the southwest horizon.

Significantly, this alignment with Menkaure/Mintaka and the Menkaure Queens with the three Belt stars occurs at a very unique moment in the precessional "pendulum swing" of the Orion's Belt stars. This quite unique moment—marked by the 212 degrees Menkaure/Mintaka alignment and corroborated by the arrangement of Menkaure's Queen's Pyramids—is the very moment the Belt stars stop and reverse their precessional direction to begin their long 13,000 year journey back to their point of origin. This unique moment, circa 10,500 BCE is known as the precessional *minimum* culmination. After another 13,000 years drifting in the opposite direction, the Belt stars will reach their precessional *maximum* culmination, circa 2500 CE, and so the precessional pendulum swing of the Belt stars continues—forever.

There is yet more evidence to present regarding this quite extraordinary relationship between the Giza pyramids and the Orion's Belt stars. As already mentioned, the Belt stars swing across the sky like a clock's pendulum, moving imperceptibly slowly between minimum and maximum precessional culminations. We have already observed how the Menkaure Queens have been placed as "markers" to indicate the precessional minimum culmination of the Orion's Belt stars when these stars were aligned in horizontal fashion close to the southwest horizon. This then begs the obvious question—how and where will the Belt stars be aligned when rising at their precessional *maximum* culmination on the future date of circa 2500 CE?

Consulting astronomical software, we find that the Belt stars will rise at maximum culmination on the eastern horizon rotated 90 degrees (perpendicular) to the stars setting at minimum culmination. Astonishingly, this is precisely the arrangement we find that the other three Queen's Pyramids beside Khufu's Great Pyramid have been placed in.

It is also worth noting that in considering the Queen's Pyramids as "precessional markers," this might help explain the curious absence of any queen's pyramids of Khafre, a pharaoh who reputedly had five queens. Since only the precessional max and min culminations need be demonstrated (there is little need to mark any intermediate point), thus the mysterious absence of Khafre's queen's pyramids is simply and logically explained.

In conclusion, the two sets of Queen's Pyramids at Giza demonstrate the precessional minimum (setting) and maximum (rising) culminations of the Orion's Belt stars, a process that takes some 13,000 years just to complete one half-cycle. That such astronomical information is exhibited to us in plain view at Giza is truly extraordinary, and provides strong corroboration to the Orion Geo-Stellar Fingerprint, which—as we have seen—defines the base dimensions of the three main pyramids at Giza. So, in three distinct sets of structures at Giza, the constellation of Orion is clearly and unequivocally implicated, thus providing corroboration to the Orion Geo-Stellar Fingerprint.

Yet the Orion-Giza story isn't quite over yet—there's more.

One of the main points of contention that opponents of the

Orion-Giza hypothesis leveled at Bauval's work was the fact that the Belt stars do not present a perfect, center-to-center correlation with the three Giza pyramid centers. The error, however, is very small and, indeed, it would seem that this minor error was quite *intentional*. We can deduce this from the fact that the Giza pyramids match the Orion Geo-Stellar Fingerprint so accurately. Such an accurate match could only have been created from having made a near perfect observation and recording of the belt star asterism.

Furthermore, when we place the Belt stars over the pyramid centers, we find that the center star (Alnilam) is slightly offset from its pyramid center. However, when we circumscribe the pyramids precisely within a tight circle using the three most extreme pyramid corners to define the circle's circumference, we find that the center of this Great Giza Circle (GGC) lies almost perfectly on the center of Alnilam.

It seems, then, that the ancient designers of Giza measured and placed the belt star asterism with high accuracy on the ground at Giza, but for some reason decided to place Khafre's Pyramid slightly offset from the Alnilam/GGC center.

Intriguingly, independent researchers Scott Sacharczyk and Rob Miller have found that the offset between Khafre center and the center of the Great Giza Circle measures precisely 44 by 14 cubits, which represents an approximation of *pi*: 44/14 (or reduced to 22/7) = 3.14285714, thus presenting the possibility that the G2 offset was deliberately created to encode the pi formula and, by extension, the Great Giza Circle.

Khafre's Pyramid has two entrances, a unique feature that scholars such as Miroslav Verner, Amelia Edwards (1831–1892), Ahmed Fakhry (1905–1973), and Vito Maragioglio and Celeste Rinaldi (per their 1963–1975 study) have regarded as perhaps indicating that the location of Khafre's Pyramid was changed from an earlier planned location, which would have placed it slightly further north and east of where it presently stands. This view agrees with the findings of the Orion Geo-Stellar Fingerprint and the center of the Great Giza Circle.

To summarize, through the arrangement and dimensions of the various structures at Giza we are presented with multiple pieces of quite diverse evidence, all pointing to a clear connection of the Giza pyramids with the Belt stars of the Orion constellation. With this mathematical and astronomical evidence—together with the mass of cultural evidence previously cited by Bauval, Gilbert, and Hancock supporting such a correlation—for Egyptologists to continue to reject that such an association was fully intended by the ancient designers must surely now be considered an untenable position.

A sufficient body of evidence now exists to allow us to safely conclude that the structures at Giza were constructed with reference to a master plan. The evidence presented here in support of this view may not *conclusively* prove an Orion-Giza association, but it is probably fair to say that if it is weight of evidence that counts, then we must consider that the Orion-Giza association now has sufficient evidence to allow us to consider the hypothesis as proven *beyond reasonable doubt*.

The Giza "Clock of Ages"

In acceptance of the Orion association with the Giza pyramids, what remains now is to consider the "why" question: Why did the ancient designers expend so much blood, sweat, and tears to create this great astronomical clock, indicating the past date of circa 10,500 BCE and the future date of circa 2500 CE? Why are the ancient designers indicating this 13,000-year half-cycle of time in a series of monuments that could easily have been built to survive such a lengthy period of time—and more? And is it perhaps significant that around 10,500 BCE—the first date indicated in the astronomical clock—the last ice age came to an abrupt and somewhat dramatic end with all manner of animal and plant species becoming extinct at that time? What event could have occurred on the Earth to bring about such dire global consequences?

Whatever inspired the design and later construction of this quite monumental astronomical "Clock of Ages" at Giza, it seems evidently clear that it must have been of vital importance. The two sets of Queen's Pyramids at Giza present the Belt stars of Orion aligned in two different ways, indicating two different dates: a horizontal alignment (minimum culmination indicating circa 10,500 BCE) and a perpendicular alignment (maximum culmination indicating circa 2500 CE). Our civilization ignores the dates indicated to us with these two alignments—these Orion "signs"—at our peril.

NINE

The Cygnus Mystery

HAVE COSMIC RAYS AFFECTED HUMAN EVOLUTION?

ANDREW COLLINS

Did cosmic rays have a hand in effecting shifts in human evolution, from Paleolithic times through to the modern day? Has this helped determine not only our physique and behavior but also our creativity and consciousness? These are wild notions, yet they are suddenly beginning to appeal to mainstream scientists and astronomers. Indeed, as long ago as 1973, American astronomer and science writer Carl Sagan wrote in his book *The Cosmic Connection* that human evolution was the result of incoming cosmic rays from some distant neutron star, demonstrating how we are right to think of ourselves as part of a greater whole at one with the cosmos.

Yet is this correct? Is Charles Darwin's theory that evolution is caused merely through survival of the fittest, and the process of natural selection, somehow flawed? The idea of cosmic radiation reaching Earth from deep space has fascinated the scientific world since its discovery, following a series of balloon ascents by Austrian physicist Victor F. Hess (1883–1964) in 1912. Then, when in the late 1920s American geneticist H. J. Muller (1890–1967) discovered that radiation (he used X-rays and later radium) was a mutagen through his work with *Drosophila* fruit flies, the subject of whether or not high-energy cosmic rays might cause changes in human DNA was voiced for the first time. Muller himself twice wrote about the subject, concluding on each occasion that the normal background fluctuation in cosmic rays reaching Earth was inadequate to explain spontaneous mutations in life forms, whatever their type. Muller was not wrong. Yet had he been privy to modern scientific data, which now confirms that at certain times in the Earth's history the solar system has been bombarded with high levels of cosmic rays, he might have thought again.

Records from the Ice

When so-called primary cosmic rays hit the upper atmosphere, almost all of them break up when they collide with nuclei of oxygen and nitrogen, the process producing a plethora of charged secondary particles. Many disintegrate in milliseconds, but others form isotopes that are preserved in everything from lake sediments to stalagmites and, more crucially, the layers of ice that accumulate each year to great depth in the Arctic and Antarctic regions.

One such isotope is Beryllium-10 (Be-10), which can be extracted from ice cores and measured to provide an accurate indication of cosmic ray activity in the upper atmosphere. It shows that over the past 100,000 years, there have been three periods when the cosmic ray flux has increased dramatically. The first was around circa 60,000 to 70,000 years ago, the second occurred approximately circa 35,000 to 40,000 years ago, and the third and last peak began around 16,000 to 17,000 years ago, and continued until around 14,000 years ago. Each spike lasted for a period of approximately 2,000 years. Similar results have been determined from a stalagmite removed from a submerged blue hole in the Bahamas. An examination of its Beryllium-10 content indicates that at various points between 45,000 and 11,000 years ago, the Earth was bombarded by twice the amount of cosmic radiation that we get today.

Where's the Cosmic Source?

The first question we must ask is where this influx of cosmic radiation might have come from. Was it really a neutron star, as Carl Sagan suggested, or could it have been another astronomical source out there in deep space? Alternatively, was there some other, more prosaic solution to this enigma? The more or less regular gaps between the spikes of Beryllium-10 activity noted in the ice cores might well indicate some kind of cyclic force in action, most obviously that of the sun. Cosmic rays are known to be partially deflected by the solar magnetic field that stretches far out into the heart of the solar system, making the rate of Beryllium-10 production in the upper atmosphere dependent on the strength of the solar field, which is itself connected with sunspot activity.

In addition to this, the sun's long-term climate cycles of 100,000, 41,000, and 23,000 years, first noted by Serbian geophysicist Milutin Milanković (1879–1958), must also affect the production of Beryllium-10 for similar reasons, such as the influence of the solar field upon the Earth's upper atmosphere. This said, there might easily have been other factors behind the sudden increase in cosmic rays hitting the Earth, the most catastrophic being a supernova, the death of a star as it expels the last of its nuclear fuel and collapses to form a high-mass compact object, most usually a white dwarf, black hole, or neutron star.

Supernovas are thought to produce enormous bursts of cosmic rays and gamma rays, which are sent careening across space at virtually the speed of light. If such an event occurred close enough to our own solar system, then the Earth would be showered by deadly radiation. This would damage the ozone layer, causing not only many more rays to reach the surface of the planet, but also the onset of high levels of UV radiation from the sun. More conservatively, catastrophists suggest that cosmic rays from a close supernova would dramatically increase cloud formation, preventing the sun from penetrating through the atmosphere, thus bringing about a sudden ice age.

Whatever the consequences of a close supernova, life on Earth would suffer mass extinctions. As terrifying a scenario as this might seem, it was the favored theory for the sudden disappearance of the dinosaurs some 65 million years ago until the discovery in 1980 of the Chicxulub impact crater in Mexico's Yucatan peninsula. This helped confirm the alternative theory that a super-sized asteroid or comet had been responsible for their extinction. Indeed, the supernova solution had been the choice of Carl Sagan and his coauthor Dr. I. S. Shklovskii, the famous Soviet astrophysicist and radio astronomer, in a book entitled *Intelligence in the Universe,* published in 1966. In fact, one wonders whether Sagan's unique view that cosmic rays have accelerated human evolution actually stemmed from his obvious fascination with the extinction of the dinosaurs.

Yet the powerful idea of a close supernova wreaking devastation on Earth during some past geological age lingers, with some catastrophists believing that it could have brought about mass extinctions during other geological epochs, for instance at the close of the Jurassic Age some 145 million years ago, as well as at the culmination of the Pleistocene Age, which coincided with the end of the last ice age some 12,000 years ago. And such scientific speculation is where it starts getting interesting, for when the high levels of Beryllium-10 were first noted in the ice cores at the beginning of the 1990s, scientists from the Cosmic Ray Council of the Soviet Academy of Sciences, working alongside a team from the University of Arizona, speculated that those around 35–40,000 years ago probably resulted from a supernova explosion.

To back up their dramatic claims, the joint Soviet-American team cited the presence of an immense formation of glowing clouds of gaseous debris—the remnants of an unimaginable supernova explosion around 150 light-years away (that's just 900 million, million miles from here) in the northern constellation of Cygnus. Had this remnant of a supernova explosion—known to astronomers as the Cygnus Veil, or Veil Nebula—been responsible for showering the Earth with cosmic rays for anything up to 2,000 years some 35,000 to 40,000 years ago? Did it bring about dramatic climatic changes and bursts of radiation that evolved humanity into what we are today?

The Emergence of Man

For whatever reason, the worldwide press coverage that resulted from this dramatic announcement of a close supernova decimating the Earth some 35,000 years ago came to nothing. Yet, thankfully, there was one person who did take notice—British anthropological writer Denis Montgomery. Having lived in Africa for many years, where anatomically modern humans emerged for the first time around 200,000 years ago, he became intrigued as to why sudden jumps in evolution occur. Was it purely spontaneous, through chemical changes in the body, or were there other exterior factors at play, such as environmental and climatic changes, nutritional variety, interbreeding, or even simple competitiveness?

Although there is ample evidence that our earliest ancestors migrated from Africa, most probably in search of new food resources, as early as 70,000 to 80,000 years ago, only tiny glimpses exist of what we were capable of achieving at this time. For instance, around 80,000 years ago, the peoples of the Republic of the Congo were making barbed bone hooks for fishing, while a community that inhabited a large cave at a place called Blombos on the southern coast of South Africa would seem to have fashioned the earliest known examples of expressive art.

These take the form of incised pieces of red ochre, showing recurring cross-hatch designs, as well as perforated snail shell beads, once strung on a cord and worn either as a necklace or a bracelet. All of these invaluable objects are thought to be approximately 75,000 years old. Then there is the recently discovered archaeological evidence from a remote mountain cave in Botswana sacred to the indigenous San bushmen. This shows that ritual activity has been occurring here in a similar manner for anything up to 70,000 years, around the time when the first migrations out of Africa are thought to have occurred. Strangely, this was also when the ice core samples tell us that there was a dramatic increase in cosmic radiation hitting the Earth, the first of three major bursts in the past 100,000 years.

Age of the Artist

Yet aside from this clear evidence of human creativity and imagination 70,000 to 80,000 years ago, it was not until the start of the Upper Paleolithic Age around 40,000 years ago that something quite dramatic started to take place. At a time coincident to when *Homo sapiens* first entered a Europe dominated by his Neanderthal cousins, there is clear evidence for the sudden emergence of a complex life style, the earliest known to humankind. It involved religious expression and practices, including detailed funerary rites, as well as magnificent new forms of art, such as the carving of animals, birds, and humans in bone and stone. It also involved, crucially, the manifestation of highly sophisticated cave art, such as the extraordinary painted galleries discovered as recently as 1994 at Chauvet in France's Ardèche region.

Occupied as early as 30,000 to 32,000 years ago, it contains images and sculptures of whole menageries of wild animals, including horses, rhinos, lions, mammoths, and bison. Alongside these are perhaps the oldest known representations of the human form anywhere in the world. These are represented by painted torsos and legs of a large-bodied woman, typical of later "Venuses" found either in statue form or as high relief in other caves, and an accompanying bison-headed figure labeled the Sorcerer, both of which are to be seen in the deepest part of the cave system.

Rapidly, hundreds of caves across western Europe became full of accomplished art forms, a tradition which lingered until approximately 17,000 years ago, when a renewed interest in sacred painting deep underground suddenly developed. This trend finally ended around 12,000 years ago when the Upper Paleolithic Age climaxed coincident to the cessation of the last ice age.

What Denis Montgomery wondered was whether, in addition to other environmental, climatic, and human factors, the increase in cosmic rays approximately 35,000 years ago (perhaps from the assumed supernova explosion which caused the creation of the Cygnus Veil) acted as a mutagen to effect sudden changes in the brain's neurological processes. This, in turn, might have brought about the enlightened age of the cave artist in western Europe. It could also explain why the Neanderthal peoples suddenly became extinct around this time, perhaps as a result of too much competition from their competitive new neighbors, the *Homo sapiens*.

Montgomery's unique ideas were privately published, and, inevitably, largely ignored by the scholarly community. Adding to his problems was the realization by astronomers during the mid-1990s that the Cygnus Veil, the nebula at the center of what Montgomery came to call the "Cygnus event," was found to be not 150 light-years away from Earth, as had previously been thought, but much further away, probably around 1,800 light-years' distance from here. At this greater distance, any supernova would have been little more than a bright light source in the northern sky, lasting for a period of several days before gradually dying away.

Doubly damning were recalculations concerning the age of the supernova event, which now appears to have occurred as recently as 5,000 to 8,000 years ago (even though some astronomers still reckon it took place much earlier, perhaps 10,000 to 15,000 years ago). Thus there was no way that the Cygnus Veil could have been responsible for the high levels of cosmic rays reaching Earth's atmosphere prior to the emergence of the first European cave artists some 32,000 years ago. So where did they come from?

Enter the Meinel Group

It would not be until 2005 that this same cosmological conundrum would again be tackled, this time by an academic think tank from Nevada. At the conference of the Theoretical Archeological Group (TAG) in Sheffield, England, held in December of that year, Dr. Aden Meinel—professor emeritus of the College of Optical Sciences at the University of Arizona and distinguished veteran of NASA's Jet Propulsion Laboratory, who in the 1980s was responsible for the launch of space telescopes such as the Hubble—told a packed audience of archaeologists and students that the high levels of Beryllium-10 in the Greenland and Antarctica ice cores indicated that cosmic rays were responsible for germ-line mutations in both animals and human life around 35,000 to 40,000 years ago. This, he reported, had been the reason for the emergence of Cro-Magnon man in western Europe, and the sudden disappearance of the Neanderthals at the same time.

In addition to this, Meinel revealed that he and his colleagues had been able to use the ice core evidence to determine the approximate astronomical coordinates for the source of the cosmic rays.

They had pinpointed an area of the sky in the northern hemisphere, coincident to the constellation of Draco. Here they searched for a possible source of cosmic rays and settled on a planetary nebula (a mass of glowing gas and cloud) known as the Cat's Eye. This, Meinel's group proposed, was the remnants of a galactic binary system consisting of a super giant and a once-active black hole that had spewed out jets of plasma, superheated ionized gas, along its line of axis at close to the speed of light. These, he said, had penetrated thousands of light-years of space to reach the Earth around 35,000 to 40,000 years ago, causing the changes in evolution witnessed at this time.

It was a bold theory. Unfortunately, astrophysicists are unanimous in their opinion that the Cat's Eye nebula does not contain (and never has contained) a black hole able to produce cosmic rays that might reach the solar system. However, the Meinel group are sticking to their guns, and remain convinced that the Cat's Eye nebula is the source of the cosmic rays that they believe affected human evolution in Paleolithic times. Yet tellingly, before being beguiled by the beauty of the Cat's Eye nebula, the Meinel group had originally determined the direction of the cosmic ray activity detected in the ice core samples as coming not from Draco, *but from neighboring Cygnus, the constellation of the swan*. So had they got it wrong? Were the cosmic rays coming, as Denis Montgomery had surmised, from somewhere in Cygnus after all?

The Oldest Constellation

This is where I enter the frame. My own independent research into the emergence of primitive societies, with their unique cosmologies and religion, had revealed an inordinate interest in one particular constellation—Cygnus, the celestial bird or swan, better known today as the Northern Cross. Indeed, it features as one of the oldest known

Fig. 9.1. The well scene in the cave of Lascaux in France's Dordogne region. Located in the deepest part of the caves, the fresco dates from around 15,000 BCE and seems to show Cygnus as a bird on a pole and a bird-man falling into a trance. Image provided courtesy of Andrew Collins.

artistic representations of a constellation anywhere in the world, for it is seen on the walls of the famous Lascaux cave in southern France, which is known to have first been occupied around 17,000 years ago. Here it appears in a fresco found in the cave's deepest part, known as the well shaft, as a bird-man falling into a trance next to a charging bison and a bird on a pole. This is likely meant to represent the so-called sky-pole of the shaman used universally to enter the sky-world via a cosmic axis, located in the vicinity of the north celestial pole, or Pole Star. From around 16,000 to 13,000 BCE, this was located amid the stars of Cygnus, which even by this time would appear to have been seen as a sky-bird of some sort.

Cygnus was also very likely the inspiration behind the appearance of the Venus and Sorcerer fresco in France's Chauvet Cave, where the woman's legs and thighs might well signify an abstract representation of the region of the Milky Way known as the Dark Rift, which opens out in the Cygnus region. Throughout the ages this region of the sky, close to where the ecliptic, the path of the sun, crosses the Milky Way, has been viewed as the vulva and womb of a sky-goddess, or Cosmic Mother, who gives birth to the sun. Sometimes the Cygnus stars are even seen to be attached to the Earth via a kind of invisible umbilical cord, showing it as a nourisher of life.

Cygnus also appears as a bird in Church Hole cave in Derbyshire's Creswell Crags alongside cave art dated to 12,800 years ago. A 12,000-year-old stone temple—the oldest anywhere in the world—at Göbekli Tepe in southeast Turkey seems aligned

Fig. 9.2. The Pre-pottery Neolithic cult centre of Göbekli Tepe near Urfa in southeast Turkey, constructed around 12,000 years ago. Did its shamanic elite inherit ideas relating to flashes of light caused by cosmic rays from the Paleolithic cave artists of Western Europe? Image provided courtesy of Andrew Collins.

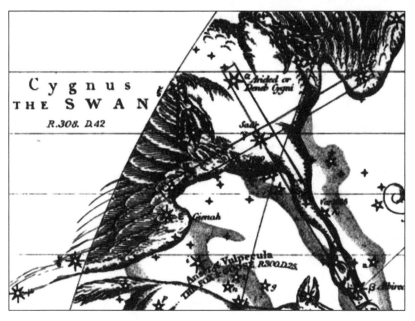

Fig. 9.3. The Cygnus constellation, shown as both a swan in flight and Cross of Christ. Author's collection.

to this same constellation. It is the same story with ancient stone and earthen structures worldwide—from the bird effigy mounds of North America to the Olmec centers of Mexico, the Incan sacred city of Cuzco, the Egyptian Pyramids of Giza, the Hindu temples of India, to the Irish and British megalithic stone complexes of Newgrange and Avebury—all of which seem to reflect an age-old interest in Cygnus.

Putting aside more obvious astronomical reasons why our ancestors might have favored this particular constellation in their religious beliefs and practices, I searched for other reasons why it was depicted deep underground by the cave artists of the Upper Paleolithic Age. In the knowledge that the work of South African anthropologist and rock art specialist David Lewis-Williams had determined that much prehistoric cave art was inspired by shamans in mind-altered states, I wondered whether the stars of Cygnus had come to be associated with religious experiences deep underground, where their most sacred cave art was executed.

Children of the Swan

I searched for answers and found that in the early to mid-1980s, underground particle detectors in different parts of the world began detecting incoming cosmic rays from deep space. Since they came in cycles of exactly 4.79 hours, the source was easily determined, for this same cycle had already been recorded in connection with other forms of electromagnetic radiation inbound from an object called Cygnus X-3, located some 37,000 light-years away in the heart of the Cygnus constellation. So inexplicable were these peculiar, neutrally charged, strongly interacting particles, resonating at some of the highest energies ever detected, that they were quickly dubbed "cygnons," later changed to "cygnets," meaning "children of the swan." This amazing data led to controversial claims that Cygnus X-3 was the first identified cosmic particle accelerator in the galaxy.

Cygnus X-3 is a binary system composed of a dying Wolf-Rayet star that feeds a close proximity neutron star (or, some suggest, a black hole or strange quark star) producing streams of superheated plasma (ionized gas). This ejecta is shot out at relativistic speeds, very close to the speed of light, along its line of axis, causing jets of debris that reach out into the local stellar medium for tens of light-years of distance. These unimaginable beams, like cosmic searchlights, are held together by magnetic sheaths that produce powerful particle acceleration in a variety of frequency ranges, including X-rays, infrared, radio waves, and gamma rays. This is not uncommon in so-called compact stars, like black holes or neutron stars, but in 2000, astrophysicists announced that Cygnus X-3 might well be the galaxy's first blazar, a term

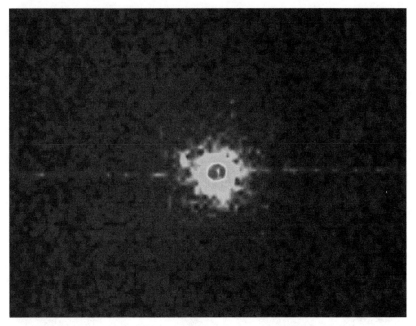

Fig. 9.4. Cygnus X-3 as photographed by the Chandra X-ray platform in 2000. Is this the source of cosmic rays influencing life on Earth? Image provided courtesy of NASA/SRON/MPE.

used when one of a deep space object's twin plasma jets is aligned toward our solar system (in other words, it is pointing straight at us). Other blazars have been identified outside of the galaxy, but this is the first time that one has been suspected to exist in our own backyard, so to speak.

What this means is that we are looking straight down the barrel of the most dangerous cosmic cannon in the galaxy, and have been, according to astrophysicists, for anything up to 700,000 years. The significance of this is that the ejecta produced by such jets could easily be responsible for increased levels of cosmic rays reaching the Earth. This includes Cygnus X-3's unique cygnet particles that, being neutrally charged, reach the Earth directly from source, and are able to penetrate deep underground, which

Fig. 9.5. NASA reconstruction of Cygnus X-3, showing its two components—one a compact object, either a black hole or neutron star, with accretion disc and twin relativistic jets, and the other a large Wolf Rayet star. Although over 30,000 light-years away, can its actions affect life on Earth? Image provided courtesy of Walt Feimer, NASA/Goddard Space Flight Center.

Fig. 9.6. Suspected black hole M87 as photographed by Hubble. Notice the relativistic jet spewing out superheated plasma at close to the speed of light. Is one of Cygnus X-3's relativistic jets aimed straight at the Earth? Image provided courtesy of NASA and The Hubble Heritage Team (STScI/AURA).

is something that cosmic rays are usually unable to do, since they are positively charged and break up before they reach the surface of the planet (neutrinos, which are negatively charged cosmic particles with almost no mass, pass through the Earth all the time without affecting anything). More important, recent findings by Japanese and Chinese scientists using data from a facility in Tibet have shown that there is, even today, a huge excess of high energy cosmic rays coming from a point in the Cygnus constellation close to the astronomical coordinates of Cygnus X-3.

Seeing the Light

This staggering scenario might well explain why our ancestors came to recognize the celestial swan as so important to their religious mind-set, since there is every reason to conclude that ancient shamans who achieved altered states of consciousness in deep cave settings, most obviously using hallucinogens, somehow became aware of the effect Cygnus was having on their lives. This might seem impossible. However, there is every chance that they would have been able to see the disintegration underground of cygnet particles, through a process known as Cherenkov radiation, which allows decaying cosmic rays to be seen as ghostly flashes of white or blue-white light as they pass through the aqueous part of the eye.

Astronauts first discovered this phenomenon in 1968 when they were aboard the *Apollo 11* spacecraft. As they tried to get off to sleep, they reported seeing "flashes and streaks" before their eyes (scientifically known as "phosphenes"). This occurred with their eyes open or closed, something that recurred during future space missions, prompting a series of onboard experiments that proved that they were being caused by cosmic rays passing through the hull of the space vessel.

Like Looking into the Universe

Hungarian-born Cornelius A. Tobias (1918–2000), a founding member of Lawrence Berkeley National Laboratory's Donner Laboratory and an expert on space biology, had earlier predicted the level of cosmic radiation that future astronauts would be exposed to, and even described its potential effects. More significantly, he predicted that they would also see flashes of light before their eyes. To test his hypothesis, he devised a unique, but very dangerous, experiment. He decided to expose himself to subatomic particles produced by UC Berkeley's Bevalac particle accelerator, which has been described as a veritable cosmic ray factory.

Part of its function is to rip away electrons from heavy elements including iron, and then focus the nuclei into a beam of particles, which are then accelerated to virtually the speed of light, like the relativistic jets produced by compact objects such as black holes and neutron stars. Tobias quite literally stuck his head in the flow of the particles and observed something almost unique on Earth. "You see visual flashes," he recalled, shortly before his death in 2000. "It is an exhilarating sensation. It is as though you are looking into the universe itself." Tobias repeated the experiment, even introducing his colleagues to the experience, until the tests were finally discontinued on health grounds.

Although it is suspected that these subatomic particles pass through the aqueous part of the eye, more recent research by Livio Narici of the National Institute for Nuclear Physics in Rome has suggested that some cosmic rays might hit the brain directly, causing not only the sensation of light flashes but also other seemingly external effects such as odd smells. Narici believes that phosphenes can be created when the particles hit the visual cortex, along with the olfactory bulb (causing smells) and possibly even the central nervous system. If so, then they could pose a potential health hazard to astronauts on future space missions, especially if phosphenes, or indeed other forms of sensory phenomena, occur at crucial moments such as when maneuvering vehicles to land. Narici has now been granted permission to conduct experiments with astronauts onboard the International Space Station (ISS) under the project name of ALTEA.

A helmet-shaped multisensor device will be worn by volunteers for an hour at a time to monitor the passage of incoming cosmic rays. At the same time, the astronauts will log when they see light flashes or trails. Hopefully, the two will coincide, telling Narici and his team where exactly the particles are hitting.

Under normal circumstances, we do not see cosmic rays down here on Earth, since we rarely experience total darkness. Moreover, the vast majority of cosmic rays hitting the Earth are broken up in the upper atmosphere, and fall as showers of harmless secondary particles. Even if we were to experience total darkness at ground level, so many other forms of environmental radiation, whether natural or industrially produced, might additionally cause phosphenes, meaning that any produced specifically by incoming cosmic rays would be lost in the process.

Virtually all forms of environmental radiation are shielded out deep underground. This is the reason why particle detectors and accelerators are located deep underground in mines, or off the side of tunnels inside mountains. Only here can they be sheltered from extraneous radiation, including stray cosmic rays penetrating the overhead rock. This, then, is what makes the cygnets from Cygnus X-3 so unique. Only strong, neutral particles of this kind are able to penetrate depths of hundreds of feet, before finally breaking up to cause secondary particles known as muons. It is for this reason alone that they were first detected by underground facilities, which at the time were attempting to witness the decay of a subatomic particle known as the proton. They included the Soudan underground mine facility in Minnesota and the NUSEX experiment in the Mont Blanc facility in southern France.

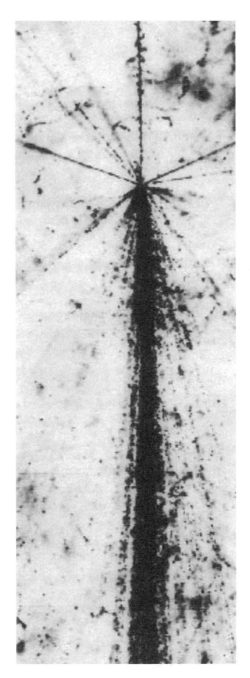

Fig. 9.7. A micrograph showing a cosmic ray impact.
Image provided courtesy of Andrew Collins.

Thus a shaman in the total darkness of a deep cave setting would be troubled only by incoming cosmic rays from just a very few deep space sources, the most likely being Cygnus X-3 (although some facilities have detected a weaker cosmic particle inbound from a source designated Hercules X-1). Intriguingly, I have spoken to a woman whose father worked in the Soudan underground mine before the particle detection facility was built at the beginning of the 1980s. She claims that he would experience unaccountable flashes of light in the total darkness. I have heard similar stories from those who have spent long periods inside deep caves and mines in Britain, including the Cheddar Caves in southern England and an abandoned slate mine beneath a mountain at Dinas, West Wales. All report unaccountable flashes of light in such environments. Even though geologically produced radiation, such as radon gas, might account for the production of some phosphenes underground, there is every chance that some are the result of incoming cosmic rays.

Closer to God

Is this, then, what our Paleolithic ancestors also experienced deep underground—flashes of light caused by the passage of cosmic rays inbound from sources such as Cygnus X-3? Would they have interpreted such experiences in a religious context? I think the answer is going to be yes. On the isolated peninsular of Mount Athos in southern Greece, ascetic monks adhering to a form of religious devotion known as hesychasm, also called omphaloscopy, or "navel-gazing," have for centuries repaired to the darkness of caves on the mountain to witness the divine light and glory of God, which is compared with that experienced by the disciples on Mount Tabor at the time of the Transfiguration of Jesus. It is a process that involves a period of deep meditation and contemplation that can last for several days, perhaps even longer. Yet eventually, I have been told firsthand, they witness flashes of light, which are interpreted as the Light of God. When this occurs it is often accompanied by visions of Jesus, the Virgin Mary, and the Christian saints—the whole process is said to bring them closer to God.

No one knows the origins of hesychasm, although it is thought to have been introduced to Mount Athos, its principal center, from Cappadocia in eastern Turkey, the home of the earliest Christian Fathers, and some of the most austere Christian ascetics. Such religious practices most probably came originally from shamanically based notions that had their genesis among the earliest Neolithic communities of central and eastern Turkey, including the city of Çatal Hüyük near Konya and, before that, the Pre-Pottery Neolithic cult centers of Göbekli Tepe, circa 10,000 BCE, and Nevali Cori, circa 8400 BCE.

Although apparently unconnected to the term omphaloscopy (which is a derisory term), it should be pointed out that the navel (the Greek *omphalos*) is a primary symbol of the axis mundi, the axis of the Earth, thought to be linked via an invisible sky-pole or umbilical cord to a cosmic axis located in the vicinity of the North Star. Curiously, the name Göbekli Tepe means in Turkish the "hill of the navel," suggesting that it was once thought to be an axis mundi in its own right.

Feels Like Heaven

I believe it highly probable that the mental discipline of "seeing the light," or in other words, witnessing flashes of light in deep cave settings, is something that was first experienced by Paleolithic shamans, in Europe especially. It was then inherited by the shamanic elite responsible for the construction of subsurface cult centers built by the earliest Neolithic peoples. The idea of spending time in the darkness of cavelike environments was, I suspect, a key element in their beliefs and practices. Such ideas brought with them the vision of a communion with otherworldly influences that, as space biology expert Cornelius Tobias commented in connection with his own experiences of seeing the light, made you feel like "looking into the universe itself."

What we also know is that flashes of light produced in deep cave settings by cosmic rays coming from Cygnus X-3 will have increased and decreased in accordance with the presence overhead of the Cygnus stars, enabling the Paleolithic shamans to eventually synchronize their chthonic beliefs and practices with its cosmic ray cycle, and thus identify this astronomical region as the source of origin of their visionary experiences.

Moreover, the appearance of seemingly objective flashes in the eyes might additionally have been taken, as is the case of the hesychasts of Mount Athos, as manifestations of some kind of primal cause, triggering more complex connections with what might have been conceived of as a divine being. This scenario might well have led our ancestors to learn what science is confirming today—that life came from the stars. Indeed, the theory of panspermia, literally "life everywhere," proposes that the most primitive forms of life probably arrived on this planet having hitched a ride either on a comet, meteor, or asteroid. In many ways, it is the origin of the belief that we come from heaven, and will return there in death.

In Europe, it was the swan that was said to carry the souls of the dead into the next world, which was located in the north, the direction in which swans migrated to their breeding grounds each spring. In the Baltic region it was the swan that took the place of the stork in bringing babies *into* the world. It is very likely, for these reasons, that the stars of Cygnus, an ever-present northern constellation, became associated with migratory birds such as the swan. However, in regions where the swan was absent, other avians took its place.

For instance, it is clear that in the Near East, the bird of Cygnus was originally the vulture, which in early Neolithic practices at places such as Çatal Hüyük was involved in the excarnation process of denuding bodies and then, as a psychopomp (the Greek word for "soul carrier"), accompanied the soul of the deceased into the afterlife.

Cosmic Mother

In my opinion, this communion with the great unknown in deep cave settings led the ancients to celebrate the idea that we are star-stuff by teaching that the sun was periodically reborn from between the thighs of the Cosmic Mother, symbolized by the Milky Way's Great Rift which begins in the Cygnus constellation (something that might indeed be depicted in the Paleolithic Chauvet Cave in southern France). Moreover, I feel sure that at least a proportion of the cosmic rays that arguably caused mutagenic changes in DNA during Paleolithic times came from the direction of Cygnus, the location searched originally by both anthropological writer Denis Montgomery and ex-NASA scientist and astronomer Aden Meinel for a source of cosmic rays hitting the Earth.

Such an intrusion into human consciousness from a deep space object in the Cygnus region might well explain why the cosmic bird has been at the heart of religious beliefs and practices since Paleolithic times. Worldwide, there is a tradition concerning a sky-bird that either lays a cosmic egg that then becomes the universe, or gives forth a honk, or call, that brings the universe into manifestation. It is present in India, Egypt, and even in the Pacific South Seas, and in each instance the bird is said to be represented by the stars of Cygnus.

When in 1973 Carl Sagan wrote that cosmic rays were responsible for changes in human evolution, he boldly asserted that their source was most probably a distant neutron star. Today there can be little doubt that the neutron star in question is Cygnus X-3—the galaxy's first blazar, as well as the best candidate by far for at least a proportion of the cosmic radiation responsible for the acceleration of human evolution at a time when we were just beginning to emerge as modern human beings. Yet more disturbing is the fact that Cygnus X-3 is out there now, its cosmic gun barrel trained toward the solar system, ever ready to release volleys of cosmic debris and other types of electromagnetic radiation in our direction. Astrophysicists studying Cygnus X-3 are waiting for what they describe as the "next big bang"—showers of cosmic particles on a level never seen before, and when this happens, who knows, it might well signal the commencement of the next upgrade in human evolution.

Electromagnetism and the Ancients

DID ANCIENT HUMANS HAVE KNOWLEDGE OF THE ELECTROMAGNETIC SPECTRUM?

GLENN KREISBERG

It's been suggested, at various times, that ancient humans had knowledge and use of unseen powers, forces, and energy fields. Could these unseen forces and fields consist of electromagnetic frequency waves and particle fields that make up the electromagnetic spectrum? This is not a simple question to answer.

What evidence exists, and what kind of evidence may come to light, to support such a claim? There is no question that, as it has always existed, the electromagnetic spectrum is a naturally occurring part of our environment, comprised of a continuous sequence of electromagnetic energy arranged according to wavelength or frequency, as generated by particle motion (vibrations) and pulses created from many sources. There is also no doubt that many ancient cultures had a connection with nature and natural forces that was fundamental and could only be described as intimate and profound in ways we moderns can merely attempt to comprehend.

This article will examine some of what evidence (and possible evidence) exists that suggests ancient knowledge of the electromagnetic spectrum, examine its scientific foundation and whether it can be used to form a hypothesis and, hopefully, be applied to solving this mystery.

From ancient times to today, humans have demonstrated an inherent curiosity and the desire to understand mysterious and odd phenomena, signs, and images. For the vanished cultures of Egypt, Sumer, and other early civilizations, and actually for the entire lapsed time of humankind, there remain many unsolved and unsettled images, messages, texts, tablets, artifacts, inscriptions, engravings, schemes, and phenomena that suggest a connection to unseen forces.

As modern society explores the mysterious meanings of certain universal cultural myths and symbols, so, too, may have humans from earlier civilizations, who repeated and venerated various motifs throughout time and traditions. The origin and meaning of these mysterious symbols may, in fact, have remained unknown even to the ancient cultures that utilized them, the ancients knowing only that certain signs and symbols were important clues to even more ancient lost knowledge and powers.

It has been noted by many that the designs and motifs of ancient architecture often reflect—and in many ways try to mimic—the patterns, signs, and signals found to occur naturally in our environment. Most significant, for the purposes of this report, are the many variations of the basic waveforms, be it sine wave, sawtooth wave, box wave, or the endless variety of spirals and waveforms that adorn ancient cave walls, temples, and structures, and that appear in architecture, scrolls, tablets, and inscriptions throughout the ancient world.

Southeastern Native American cultures dating back 20,000 to 25,000 years extensively used waveform symbols for ornamentation on nearly all handmade items and wares, such as pottery and textiles. The variety is nearly endless. As I have mentioned, the electromagnetic spectrum exists naturally, occurring as a part of our natural environment. And again, acknowledging the ancients' intimacy and interdependency with nature, it would not be surprising if they possessed some knowledge of this naturally occurring "tool."

The ancient cultures of this world are known to have identified and utilized the forces of nature to their benefit, including water, fire, wind, and sound. Are we to believe that mankind is only now, in the past century, exploiting the waves and frequencies of the electromagnetic spectrum for the first time? I'm not so sure. And perhaps more important, if ancient cultures did possess this knowledge, where did it come from, and how was it processed?

A thorough examination of this subject must begin with

Fig. 10.1. Mayan Pyramid of Kukulkan at Chichén Itzá has a sawtooth waveform built into its architecture. Photograph courtesy of SacredSites.com.

tracking the breakthroughs and discoveries that have occurred throughout history and that have led to the concepts and principles that make up modern electromagnetic theory.

Electromagnetic radiation has been around since the birth of the universe; light is its most familiar form. Electric and magnetic fields are part of the spectrum of electromagnetic radiation, which extends from static electric and magnetic fields, through radio frequency and infrared radiation, to X-rays.

From written history it appears that many of the concepts now familiar in electromagnetic theory were explored and developed during a time when many modern high-tech investigative and detection tools and methods did not exist.

But is it possible that the ability to manipulate the particles and waves of the electromagnetic spectrum was discovered and developed even earlier than written history suggests? Could it be that many of the symbols, images, architectures, and myths of ancient cultures are representations reflecting the possession of such knowledge?

Observing what he believes to be power plant plans on ancient tapestry, and microprocessor design in Egyptian temple layout, Slovakian researcher Dr. Pavel Smutny has written, "Maybe it is unusual and surprising, but in ornaments on old carpets are woven-in schemes, and principle plans of advanced technologies, which come from vanished cultures and thousands-year-old civilizations. These residues are probably the last ones which can help revive forgotten, very sophisticated technologies and methods for exploitation of natural electrostatic energy sources."*

On the Egyptian temples, Smutny continues, "If we see plans for the Valley temple at [the] Sphinx, or of the mortuary temple at Chaphre's Pyramid, or the mortuary temple at Menkaure Pyramid, or also of Osereion from Abydos, to a person familiar with the

---

*From *Microwave, Optics, Acoustics, and Ancient Egyptian Temples,* by Dr. Pavel Smutny. http://farshores.org/moae.htm.

Fig. 10.2. Photograph showing Neolithic "lobe" construction of the Temple of Mnajdra on Malta. Photograph courtesy of SacredSites.com.

basics of computer techniques, or even better to [a] person experienced with construction of microwave circuits in bands above one gigahertz (GHz), he will tell you that these plans are schemes of PCBs (boards for electronic circuits)."

Smutny further notes, "These plans look like schemes of electronic circuits working on quite high frequencies above 1 GHz, with use of so-called components with distributed (spread) parameters. These components are of such sort, that the top conductive layer is formed to various shapes, which work as inductors, capacitors, resistors, filters, resonators, or have similar functions. Advantages of so produced circuits are very good functions and low losses on high frequencies. Not so different from classical electronic circuits with use of discrete components as began to be produced in the 1980s, in at that time superior high frequency devices."

If, as Smutny suggests, these tapestry patterns and temple layouts are representations of fragmentary remnants from long-lost science, what could have been the original source, other than a previously existing technologically advanced civilization? Of course more evidence is needed before any solid conclusions can be drawn.

Turning his attention to the ancient temples of Malta, Smutny explains, "complexes were used probably as generators of high frequency acoustic waves. Purposes were (maybe) to arrange a communication channel among various islands."[*]

Legends abound in Malta of sirens (acoustically) tempting or deafening seafarers. Sirens were persons, possibly the best singers with the strongest voices, namely obese or thick women, such as some modern opera stars. Many statues of just such women have been discovered everywhere on Malta, including in temples, standing on pedestals, and in the middle of complexes. Smutny speculates, "their singing modulated low frequency signals, which were generated on opposite ends of temples (in windows and in doors) simply with a bell, or with vibrating metal plates, or even with a strong wind drafted through a wall opening."

To support his theory, Smutny points to the massive blocks used for construction of the complexes, noting that they are a very good resonant material. Smutny suggests a reconstruction experiment to prove functionality of the megalithic complexes, in which the roofs would be intact (buildings are currently without ceilings), and models made from concrete, or from stone in suitable scale. For testing, acoustical or ultrasound frequencies should be utilized, where wavelengths are diminished to scale.

As a radio frequency engineer, I find these possibilities fascinating, and while pondering them, I had a bit of a revelation: Namely, a bird's-eye view of the temple structure and configurations in Malta in many ways closely resembles an antenna propagation pattern,

---

[*]From *Acoustics in Temple Complexes on Malta's Islands,* by Dr. Pavel Smutny.

Fig. 10.3. The multichambered oval configuration of the Maltese temples allowed signals formed from groups of air particles, before output, to be amplified in a second parallel opposite oval spaces of the temple. This would be analogous with a resonator. Photograph courtesy of SacredSites.com.

with its main beam lobe, side lobes, and nulls (between the lobes).

There are an infinite variety of patterns that can be generated by antennas, depending on the beam width desired, distance between the antenna poles, and the frequency being transmitted. However, nearly all antenna patterns have similar characteristics, consisting of main beam lobe and side lobes, as well as nulls and back lobes.

I believe *lobe* was a word Graham Hancock specifically used in his book *Underworld: The Mysterious Origins of Civilization* to describe the temple structures on Malta. When Hancock mentioned the unusual acoustics he encountered in the Hypogeum on Malta, I began researching anything that referred to those acoustics or a similar effect.

In Joseph S. Ellul's 1988 publication *Malta's Prediluvian Culture at the Stone-Age Temples,* the author describes, "Before going into the entrance, just have a look at the lintel and at the long blocks lying alongside and on top of it. Most of them, including the lintel itself, which measures about nine feet square and eighteen inches thick, were scattered about in this courtyard. Almost all of these blocks had been thrown over into this yard for a distance of about twenty feet. The lintel fell on its right corner and was splintered in the fall." Interestingly, Ellul continues, "That lintel was found resting some thirty feet to the east of the doorway, lying on some rubbish and with one of the long horizontal slabs on top of it. It happened that the upper stone had certain acoustic characteristics and was positioned so as to be almost perfectly balanced, and whenever one hit it with another stone, it rang like a bell." That peculiarity earned it the nickname of "The Bell of Ħaġar Qim" but, during works carried out in the 1950s, these various blocks were put back in their present places on top of the front wall, and "The Bell of Ħaġar Qim" was no more.

Odds are, of course, that this is all simply coincidence, but it poses the interesting question: Exactly what is the unique and specific temple design layout of Ħaġar Qim based upon? What was the impetus and influence that dictated the architectural configuration so long ago?

I design digital wireless networks utilizing the electromagnetic spectrum, transmitters, receivers, modulators, and antenna. I also do propagation prediction modeling using powerful computer programs. The program creates a model showing how a location will perform using different antenna, elevation, azimuths, and power settings. Sites are chosen based on how well the location propagates a signal.

Again, when Hancock mentions in *Underworld* how the land gently dropped away for a great distance from a temple location, I couldn't help think how ideal that is for signal propagation when designing sites that can transmit great distances. Is it possible that some ancient temple locations were chosen for similar reasons? If the temple location were part of some kind of communications network utilizing the electromagnetic spectrum, I'd say they were.

Is it possible that a correlation exists between the Maltese temple architecture and antenna propagation patterns? Perhaps, but much more research and investigation is needed before any firm conclusions can be drawn. Certainly it can be said that the temple layout seems to conform strikingly to the characteristics of antenna propagation patterns.

If a connection were proven to exist, what would be the implications? Certainly, a radical rethinking of the level of technological sophistication that may have existed in ancient civilization, not to mention the question of how this knowledge came to the builders of the Maltese temples.

Let's assume, for a moment, that the ancient builders had knowledge and understanding of wave propagation. Perhaps early post-Flood cultures possessed the knowledge and understood the purpose of unseen forces but lacked the functionality. If it was known and handed down through generations that a universal force existed that could be controlled for the benefit of all humans, would not every effort be made, no matter how futile the attempt, to resurrect and re-create that functionality? This in itself could account for the existence of many megalithic structures, about the many aspects and purpose of which humankind has puzzled over since their very discoveries.

Another example of anomalous science, which includes a possible electromagnetic spectrum connection, is the Celtic round towers of Ireland. These fascinating structures, the tallest measuring 111.5 feet (34 meters), once numbered over a hundred. Many still stand today, and dot the Irish landscape. Based on the fact that nearly every tower is at the site of a known Celtic church dating from the fifth to twelfth centuries CE, scholars have suggested that the most probable construction period was between the seventh and tenth centuries CE. Initially, each of the towers were freestanding structures, but later, other buildings, primarily churches and monastic foundations, were constructed around some of the towers.

In his book, *Irish Round Towers,* Lennox Barrow states,

It is remarkable how little the main dimensions vary. In the great majority of towers the circumference at the base lies between 45.9–55.8 feet (14–17 meters) and the thickness of the wall at the lowest point at which it can be measured varies from 2.9–4.6 feet (0.9–14 meters). Doorways, windows, story heights, and diameters also follow clearly defined patterns, and we may well conclude that most of the towers were the work of teams of builders who moved from one monastery to another using standard designs. Most doorways are raised 4.9–14.8 feet (1.5–4.5 meters) above the ground. This is usually explained as being for security, to enable the monks to take refuge inside the towers

Fig. 10.4. View of Irish round tower with conical cap intact. Photograph courtesy of SacredSites.com.

during times when Viking raiders or bandits were attacking the monasteries.

There is probably some truth in this theory, but it has also been noted that it's possible that the stability of the tower had as much to do with the door heights as anything else. Basic physics dictates that the higher you could build before making an opening in the wall, the stronger the base would be. Very often the towers were filled in, even as high as the doorways.

This idea that the round towers were erected and used primarily as watchtowers and places of protection is strongly debated by the American scientist Philip Callahan. In his book, *Ancient Mysteries, Modern Visions,* Callahan discusses research, which indicates that the round towers may have been designed, constructed, and utilized as huge resonant systems for collecting and storing yard-long wavelengths of magnetic and electromagnetic energy coming from the earth and skies.

Based on fascinating studies of the forms of insect antenna and their capacity to resonate to micrometer-long electromagnetic waves, Professor Callahan suggests that the Irish round towers (and similarly shaped religious structures throughout the ancient world) were human-made antennae, which collected and transmitted subtle magnetic radiation from the sun and passed it on to monks meditating in the tower and plants growing around the tower's base. The round towers were able to function in this way because of their form and also because of their materials of construction.

Of the sixty-five towers that remain as ruins, twenty-five were built of limestone, thirteen of iron-rich, red sandstone, and the rest of basalt, clay slate, or granite—all of these minerals having paramagnetic properties that can thus act as magnetic antennae and energy conductors. Callahan further states that the mysterious fact of various towers being filled with rubble for portions of their interiors was not random, but rather may have been a method of "tuning" the tower antenna so that it more precisely resonated with various specific frequencies.

Equally intriguing is that Callahan shows that the seemingly random geographical arrangement of the round towers throughout the Irish countryside actually mirrors the positions of the stars in the northern sky during the time of winter solstice.

Archaeological excavations at the bases of the towers have revealed that many towers were erected upon the tops of much older graves, and it is known that many of the tower sites were considered sacred places long before the arrival of Christianity in Ireland. These facts point to the likelihood that the ancient Irish, like the Egyptians, the Mayans, and many other archaic cultures, understood there to be an energetic resonance between specific terrestrial locations and different celestial bodies. All across the Irish countryside particular locations were chosen, precisely designed structures were erected to gather and store various energies, and a tradition of human spiritual use of the sites arose over the millennia. While many of the round towers are now crumbling and their antenna function may no longer be operative, many visitors feel that a field of energy still permeates the sites today.

In another article, *The Mysterious Round Towers of Ireland: Low Energy Radio in Nature,* from *The Explorer's Journal,* Callahan gives further details of his discoveries: "Most books will tell you that the towers were places of refuge for the monks to hide from Vikings raiding Ireland. They were, no doubt, bell towers and lookouts for approaching raiders, but the speculations that monks escaped raiders, who no doubt knew how to smoke bees out of hives or climb the nine to fifteen feet to the door, borders on the ludicrous. Round towers are perfectly designed to be totally useless for hiding people or church treasures."

Callahan continues, "Another strange thing about the towers

is the dirt that fills the base below the high doors. Each door has a different level of dirt filling the base, as if they were 'tuned' like a pipe organ. I had long postulated that the towers were powerful amplifiers of radio resonance from the atmosphere generated by lightning flashes around the world." Upon further investigation, Callahan concluded, "The round towers proved to be powerful amplifiers in the alpha brain wave region, 2 to 24 Hz, in the electrical anesthesia region, 1,000 to 3,000 Hz, and the electronic induction heating region, 5,000 Hz to 1,000 kHz. . . . It is fascinating that just above the surface of the ground to about two to four feet up there is a null of atmospheric frequencies that get stronger and stronger until at nine to fifteen feet above the surface they are extremely strong."

Were Irish monks aware of these properties, and thus prompted to build their high doors? At every tower Callahan measured, there was a direct correlation between the height of the tower door and the strength of the waves. That the highly amplified waves occur in the meditative and electrical anesthesia portion of the electromagnetic spectrum is of utmost significance. In 1963, G. Walter researched brain EEG waves from 0.5 to 3 Hz (Delta region) and found anti-infectious effects.

It's interesting to note that recent research, including experiments conducted at UCLA by professor Valery Hunt, has shown some positive effect of electromagnetic radiation fields. Subjects who were shielded from all electromagnetic fields reported experiencing a mental breakdown, their brain no longer able to function normally. Remove the electromagnetic shielding and "normal" brain function returns. When those subjects were exposed to intensified electromagnetic fields, they reported a profound feeling of well-being. Others exposed to electromagnetic fields under controlled conditions reported hallucinatory visions similar to those reportedly experienced by shamans or individuals in an altered state of consciousness, under the influence of mind-altering substances.

Additionally, research conducted in the 1980s by Dr. Michael Persinger at Laurentian University in Canada reported that subjects whose brains were bathed in low energy electromagnetic fields while wearing a modified helmet, which Persinger designed, experienced a profound, otherworldly presence during the research sessions. Over 80 percent of the subjects involved in the experiments reported experiencing the effect while under the influence of the induced and enhanced electromagnetic field.

Perhaps also related to the Irish stone towers, are the many anomalous ancient stone structures, chambers, and circles found throughout New England. In the book *Celtic Mysteries of New England,* authors Phillip Imbrogno and Marianne Horrigan note and measure unexplained electromagnetic anomalies at stone chamber locations. The anomalies, described as "pulses," seem to occur in a specific frequency range beginning at 16.5 MHz, with peak signal strength occurring at 1,675 MHz. The authors claim that this could be suggestive of a "doorway" or "portal" function of the chambers, with a burst of electromagnetic energy released whenever the door is opened or shut.

I would suggest that, as opposed to a "portal" or "doorway" opening to another dimension of time and space, it might be more likely that the chambers' electromagnetic anomalies could indicate a communication channel or conduit for energy, opening for use in some way.

Lastly, it's important to mention the groundbreaking work of research scientist John Burke, whose book *Seed of Knowledge, Stone of Plenty: Understanding the Lost Technology of the Ancient Megalith-Builders* documents how electromagnetic fields have a beneficial effect on crop yields when seeds have been exposed. Burke and his team tested electromagnetic hot spots at locations in the Midwest and in the northeast of the United States, including some of the stone chambers already mentioned. They located and mapped naturally occurring spikes in the Earth's electromagnetic field; locations that in many cases had long ago been identified and marked as "sacred" by local indigenous shamans. What they found was that seeds they brought to these locations and left for a time period would produce more robust yields compared to a control group of seeds left in the lab.

Is it possible that the purpose for which many of the sacred sites and megalithic structures around the world were constructed was to receive or gather, process or focus, and transmit or distribute naturally occurring electromagnetic energy? Could ancient people have been aware of the medicinal and consciousness-raising properties of electromagnetic fields? There is much evidence to suggest that they were.

The Infant Field of Radio Archaeology

I would suggest a series of experiments, conducted to scientific method standards, to test and measure both the transmit and receive propagation properties of prehistoric sites that meet specific criteria. Such tests should utilize carrier wave transmitters and spectrum-scanning receivers and software, along with spectrometers, spectrum analyzers, and oscilloscopes to determine if propagation properties are enhanced by the sacred site locations and/or site configuration and construction materials.

Someone once said that science is like a game of darts. The surface of the dartboard, albeit perhaps infinite, represents all the science there is to know. Each theory confirmed through experiment and scientific method represents a dart hitting its mark on the dartboard of science. While the surface of the board may have many darts in it, most of it, as of yet, remains untouched by a dart. And, of course, sometimes one dart thrown may knock out a previously thrown dart from its apparently secure location. In other words, there is still a great deal of true science yet to be discovered and confirmed, and perhaps some ideas that may need to be unlearned. Above all, the perspective with which we search is perhaps the key to what we discover.

An important lesson can be learned from the mechanism known as the Antikythera Device—discovered over a hundred years ago on the floor of the Mediterranean Sea, the device has been found by archaeologists to be two thousand years old. Because the device suggests that the ancient Greeks had the mechanical sophistication of eighteenth century Swiss watchmakers, historians and scholars have generally ignored it. Could this device be evidence that a period of technological devolution, as a general trend, took place over many thousands of years in fields such as mathematics, mapmaking, mechanics, and more? Can the same claim be made for wave and particle theory? Was much more known, much earlier on, than current evidence supports or the academic world is willing to accept?

Nothing less than a major shift in the approach to these mysteries needs to occur for a full debate and examination of the facts and theories to take place within the established academic community. The fact is that the Antikythera Device exists, as do many other anomalies of science. No, they don't fit into the context of established history, but to fail to consider them as part of a larger picture that is far from complete does a great disservice to the pursuit of knowledge.

The ancients may have seen the electromagnetic forces of the universe as the single common denominator of nature, of all that exists in the physical as well as nonphysical world. If so, they would have seen the electromagnetic spectrum as the medium on which all existence records itself. Learning to manipulate and harness the properties of that field would be the ultimate power, the power of the gods.

From the first time a human tossed a pebble into a pond of water, the dynamics, properties, and laws of wave propagation were evident for him to examine and test. In air as in water, these properties follow many consistent traits: Greater power creates larger waves, and waves could easily be regulated and directed. Could it be that our ancestors understood the subtle properties of electromagnetic waves and particle fields and even manipulated them, harnessing the power of nature and the power of the unseen spectrum? As if to give the game away, they placed clues for us to find. All of their rituals, their religions, cults, gods, demigods, the whole of their lives were linked to the cycles of the universe. And it is important to recognize that this is how the ancients understood the powers of the universe, through their gods. Perhaps they did not conceive of electromagnetic power as we do, but rather as god-given signs and signals, as the power of the gods, to be tapped into by humans.

ELEVEN

The Gulf of Khambhat

DOES THE CRADLE OF ANCIENT CIVILIZATION LIE OFF THE COAST OF INDIA?

S. BADRINARYAN

It was generally believed that well-organized civilizations could not have existed prior to 5500 BCE. Many were reluctant to accept the flood myths mentioned in many ancient religious writings. The recent discovery made in the Gulf of Khambhat, India, shocked many and made some sit up and watch with interest. It clearly established the existence of an ancient civilization that was submerged in the sea. The methodology adopted was novel and different, wherein advanced marine technologies and the most modern scientific applications of various disciplines were put to best use. The traditional but conservative archaeologists found it hard to accept that a major discovery could have been made by hitherto unapplied, unheard-of techniques. Some observed and understood the importance of the discovery and came out in open support.

Initially, when the side-scan sonar images of underwater structures were shown, some called it the magic of computer software. When hundreds of artifacts were collected and shown, traditional archae-ologists opined that the artifacts could have been transported by using the ancient river! Detailed scientific studies were again undertaken to prove that the artifacts are in situ. The criticism has driven us to adopt most modern technologies and scientific methodologies available in the world, which have completely sub-stantiated our findings, and the results were published as research papers in reputed international journals. Now several authors are quoting the Gulf of Khambhat work as a standard benchmark methodology for modern marine archaeological surveys and inves-tigations. The discovery has clearly established the possibility of ancient civilizations that were submerged due to flooding by ris-ing seawaters after the last ice age.

The oldest city-state civilization was thought to be Mesopotamia, dating to 5500 BCE. An extensive mature civilization also stood in the northwestern part of India adjoining Pakistan and Afghanistan. This is the well-known "Harappan" civilization, which lasted between 5300 and 2800 BCE. This includes major ruined cities like Mohenjo Daro, Harappa, and Dholavira, as well as villages, craft centers, campsites, river stations, fortified places, and ports. The cit-ies had well-lined streets, arranged in straight lines, proper drain-age, sanitary arrangements, and excellent water-conveying systems, including check dams for storing water. Usage of a variety of arti-facts, metallic objects, many types of potteries, and the construction of huge structures could not have happened overnight. So there was obviously a major missing link between the ancient hunter-gatherer group of people and the "Harappan" civilization.

In India, there were many Paleolithic, Mesolithic, and Neolithic Stone Age cultures. But none of them have any remote resemblance to the type of civilization found in the Harappan sites. It is pos-sible that the missing link between the two is either undercover or has been submerged due to major sea level rise caused by melting ice sheets. It is a well-established fact that during the last glacial maxima (ice age), the seas all over the world shrank; the sea level around 18,000 BCE was about 426.5 feet (130 meters) below the present-day sea level. So, it is logical to look for such submerged civilizations near areas surrounding the present-day Indian coastal areas, especially along the paleochannel of various rivers.

National Institute of Ocean Technology (NIOT), an undertaking of the Indian government, has been carrying out several multidisciplinary marine surveys along the Indian coastal areas for various purposes. During the course of a few geological surveys in Gujarat in the Gulf of Khambhat, NIOT came across river paleochannels in the sea. These were seen to be the extension of the present-day major rivers of the area.

In a similar marine survey, in a coastal research ship during 1999–2000 when the author was the chief scientist, several unusual frames of side-scan sonar images were encountered. These had square and rectangular features in an arranged geometric fashion, which are not expected in the marine domain. Such features are unlikely to be due to natural marine geological processes. This made the author suspect that human workmanship must have been involved. The surveys were followed up by the author in the following years, and a couple of paleochannels of old rivers were discovered in the middle of the Khambhat area under 65.6–131.2 foot (20–40 meter) water depths, at a distance of about 12.4 miles (20 kilometers) from the present-day coast.

The Gulf of Khambhat forms a funnel-shaped entrant of the Arabian Sea, sandwiched between mainland Gujarat and the Saurashtra peninsula in the west. This gulf is 83.9 miles (135 kilometers) long in a north-south direction, and is more than 62.1 miles (100 kilometers) at its widest part. It is one of the roughest and most complicated seas of the world, and covers an area of about 1,158.3

square miles (3,000 square kilometers). Several major rivers, including the Narmada, Tapi, Sabarmathi, Mahi, and the Chathranji, drain into it. It has a macro-tidal range of 39.4 feet (12 meters), and the currents are up to eight knots. The sea is often subjected to severe winds resulting in very rough conditions.

This type of turbulence churns the seabed and produces enormous quantities of silt, making the seawater brownish and turbid, with the result that it is impervious to light rays. The combined effect of these conditions makes this part of the country unfit for diving and underwater operations, and makes underwater videography impossible. Hence, only instruments operated on the principle of sound, like sonar and magnetic equipment, could work here. This includes side-scan sonar, the sub-bottom profiler, and the multibeam ecosounder, apart from the marine magnetometer.

The term "sonar" stands for "sound navigation and ranging." Sonar is a sensing strategy that measures features of an environment (or medium) by the way in which that environment transmits, reflects, and/or absorbs acoustic (sound) waves. The seabed is a little-understood environment due to its inaccessibility and resistance to large scale detailed analysis. Sonar represents a clear approach of looking at the seafloor.

Side-scan sonar uses sound waves to produce images of the seafloor. The hard areas reflect more energy and are seen as dark shades, whereas softer areas do not reflect energy as well and are represented by lighter shades. This "backscatter" is absent behind objects or features that rise above the seafloor, and are represented as white shadows in the sonar image. The dimensions of shadows are used to infer the size of the objects. The system used was a digital one, which provides high-resolution sonar images of the seafloor through advanced digital technology in the 100 and 400 kHz frequency. The unit is connected to a Differential Global Positioning

System (DGPS) for the accurate position of the survey vessel and, in turn, that of the objects.

In these surveys it was the side-scan sonar that gave excellent results supported by other systems. Initially two major paleochannels of rivers were recognized. One was over a length of 5.7 miles (9.2 kilometers) and another over 5.6 miles (9 kilometers). When these were sampled, it was seen that just below a thin marine sediment cover of a few centimeters were river alluvium and pebbles typical of terrestrial river sediments, below which typical river conglomerates were observed at depth. Such evidence clearly indicated that the area presently under water was originally dry land, over which rivers were flowing. Due to different factors, they got submerged and now lie beneath the sea.

The sonar images showed regular geometric patterns in one paleochannel over a length of 5.6 miles (9 kilometers) in the sea about 12.4 miles (20 kilometers) west of the Hazira coastal area. Associated with this on either side of the paleochannel, basementlike features in a grid pattern were observed at a water depth of 65.6–131.2 feet (20–40 meters). These resemble an urban habitation site; pitlike structures can be seen in the basement on the ocean floor. Another paleochannel over 5.7 miles (9.2 kilometers) was detected off the Suvali coastal area. Here, also, similar features were observed. In general, the basementlike features were located in a linear east-west direction on either side of the paleochannel. It is seen that these features are 16.4 by 13.1 feet (5 by 4 meters) on the eastern side, whereas the westernmost part had dimensions of 52.5 by 49.2 feet (16 by 15 meters). The habitation sites are all seen to be laid in a strict grid indicating a good sense of town planning by the ancients.

There was also evidence of water conducting systems, including canals. All these point to a properly planned township, with a high level of knowledge and practice by the ancients. The area in general is seen to be covered by sand waves, which occur above the seabed. Often these cover the dwelling, but even then the shapes could be made out. Apart from the regular sites of habitation, the side-scan sonar picked up images of several big structures. Some of these structures are as follows:

There is a rectangular 134.5 by 82 feet (41 by 25 meters) shaped depression, wherein one can see steps gradually going down to reach a depth of about 23 feet (7 meters). Surrounding this depression, there is a wall-like projection on all sites. One could observe an inlet and outlet and also a separate enclosure. This looks like a tank or bathing facility now occurring below 131.2 feet (40 meters) of sea-water. It occurs near the western periphery of the town. It resembles the Great Bath that is found in the ruins of Mohenjo Daro and Harappa, where these also occur on the western side of the town-ship. There are two divisions in the tank, which may represent sepa-rate enclosures for men and women or for socially higher and lower categories of people. There are two openings, probably for inlet and exit of water to keep the water in the tank fresh and clean.

In one of the structures one can observe a long, linear, promi-nent, and well-made basement measuring 656.2 by 147.6 feet (200 by 45 meters). It nestles on high ground, and steps approach the structure on the right corner. Inside the structure are several square-shaped, roomlike features that are 59 feet (18 meters) and higher, and that are surrounded by fortifications.

These structures resemble those of the Citadel area of Mohenjo Daro, Harappa, and Dholavira, where similar structures occur at the western extremity, again on high ground. The structure was likely an administrative building meant for supervising the civic activities of the township, or could have been a place of worship. Pieces of fossilized human and animal bones, as well as natural teeth, were recovered during sampling on the eastern side of the Citadel.

The sonar image picked up a major, dilapidated structure mea-suring 623.4 by 278.9 feet (190 by 85 meters), with spaces separated by what look like collapsed walls. In front of the structure, on the bottom side, there are several rectangular, 8.2 to 11.5 by 19.7 feet (2.5–3.5 by 6 meters) basement structures resembling minor dwellings. These could have been ancient granaries for the township, probably with a dwelling place for the workers nearby. Some fossilized food grains had been collected in the neighboring areas. The granary is a regular feature in many of the Harappan sites.

There is also a basement of a buried settlement, and it measures 242.8 by 157.5 feet (74 by 48 meters). It has regular square, rectangular, and arch shapes. The darker portions are the elevated or standout features. These indicate that there are still some constructed portions partly standing up. A few square and rectangular-shaped basements are also visible to the north of the structure.

The main structure measures 131.2 by 62.3 feet (40 by 19 meters), with wall-like dark features rising to 6.6–9.8 feet (2–3 meters) above the seabed. A series of steplike features can be seen on the right side of the structure. To one corner of the main structure, a 36 by 23 feet (11 by 7 meters) rectangular depression that looks like a small tank or pond can also be observed.

Sub-bottom profiler surveys penetrate the seabed, instead of reflecting sound waves from it *a la* side-scan sonar. The waves travel beneath the seabed in different formations in different speeds, and the instrument collects the reflection data over selected frequencies. It provides good depth information on geological features, apart from delineating any suspected buried anthropogenic structures.

Standout features, which appear to be the basement and foundations of the structure were picked up below the 656.2 by 147.6 feet (200 by 45 meters), citadel-like structure. These features were picked up at regular intervals. It is observed that the foundations have been dug up to 16.4–19.7 feet (5–6 meters) in the soil, over which broad, columnlike features have been constructed, probably to help bear the load of the huge structure above.

Below the buried settlement of the 242.8 by 157.5 feet (74 by 48 meters) structure are man-made foundations like columns that can be clearly seen emerging from below the seabed, and occur as stand-out features. Here, the foundations have been dug up to 9.8–13.1 feet (3–4 meters) deep in the soil. This type of planning and construction by ancient peoples clearly reveals that they had a very good knowledge of civil and structural engineering, wherein broader and deeper foundations were provided for bigger and heavier structures, and thinner and shallower foundations for comparatively smaller structures. Likewise, almost all of the structures, including the dwelling sites, indicate a good amount of planning and design, taking into consideration the structural aspects.

Magnetic surveys were carried out by deploying a high-resolution cesium vapor marine magnetometer. The survey was for observing magnetic signatures occurring as anomalies of subsurface magnetic bodies in the area. The instrument was capable of sensing up to 0.001 nT (nano-tesla) at 1 sample rate. The values were corrected for diurnal variation so as to remove the temporal variation in the Earth's magnetic field. The corrected magnetic field value is a result of the marine magnetic components; regional geological features are very deep-seated in origin and have depth persistence, whereas one should look for very shallow and near-seabed anomalies that do not extend in depth. Several shallow, near-surface anomalies were picked up, ranging in depth from 3.3 feet (1 meter) below surface to as much as 164 feet (50 meters).

The deep-seated anomalies are at least below 1,312.3 feet (400 meters) from the seabed surface. The near-surface anomalies are covered by top sediments in general, and are likely to be archaeological sites, to be examined and explored later. Due to these surveys, a vast area has now acquired importance for archaeological purposes—including areas that were not picked up earlier by side-scan images. These surveys therefore enhance the area of archaeological interest.

This was obviously an extensive civilization, the remains and ruins of which have since been covered by shifting shoals, sand waves, and tectonism, which are very common features in the Gulf of Khambhat.

Even though a variety of objects and artifacts were collected in settlements, doubts were expressed by some persons as to whether these could have been transported by paleochannels and might, in fact, not be in situ. To clear up such doubts, detailed geochemical analyses were carried out. Ten geological soil samples and ten artifacts were chosen from the Gulf of Khambhat area. Since trace elements like titanium, hafnium, and thorium—as well as rare earth elements (REE)—are immobile, they preserve their signature without alteration, and hence reflect their primary petrogenic character. These twenty selected representative samples were analyzed using an ICP mass spectrometer.

The leaching of light rare earth elements and a prominent European anomaly suggests there is a one-to-one match between the archaeological material and Khambhat bed sediments. The ternary and binary plots of both the materials show the clustering of all samples in one place, indicating the samples are of the same host chemistry and are in situ. In other words, the archaeological materials are not transported, but are made from locally available material only.

To substantiate the findings, detailed sampling was carried out. Since the sea condition was very rough, and the water turbid and brown, sampling was carried out in areas where side-scan images show excellent results. The samples were collected by utilizing a grab sampler, dredger, gravity corer, and vibro corer. Large numbers of samples were carefully collected, systematically numbered, and properly preserved. The artifacts collected included a variety of pottery pieces, Mesolithic stone tools, a few Paleolithic macro stone tools, beads made of semiprecious stones, brick pieces, hearth material, wattle and daub structure materials, corals, perfectly holed stones, fossilized human remains, and human teeth.

Three potsherd pieces were also unearthed. These are unfired and normally sun-dried, made of clay, and are of great antiquity. They display a crosslike object and some figurines. There is also a piece in the shape of a deer's head, and a well-turned ornamental piece with a straight hole in the center. How the ancients were able to make these in stone is still an enigma.

Four additional, important objects were unearthed. The first was a fossilized jawbone (mandible) with a natural tooth kept in front. The second object was part of a carbonized wooden log. This was obtained from the top stratigraphic column at a depth of about 30–40 centimeters below the seabed, and was sent for dating. Also unearthed were rolled objects and long linear beads that, when strung together, form a necklace. As well, linear beads made of stone were found, with holes in the middle.

A series of Microlithic tools were collected at various locations. Microlithic tools are generally characteristic of the Mesolithic period and are found between Paleolithic and Neolithic Stone Age periods. The characteristic features of Mesolithic tools are that, unlike the earlier Paleolithic stone tools, these are much smaller, normally between one to five centimeters in length, and are made of finely crafted semiprecious stones. These include quartz, chert, jasper, flint, chalcedony, agate, and corundum. About 248 such tools were collected by sampling. The tools included a baked blade with a serrated edge, point, point on flakes, lunate, scraper, cores with negative chipping, and a borer. The tools have both geometric and nongeometric forms.

The Mesolithic period of western India revealed the existence of their cultural phases, namely aceramic and ceramic. In general, the Mesolithic sites confirmed the existence of a ceramic phase in its later part, containing other materials besides potteries and Microlithic tools, like wattle and clay shreds for house construction and flooring. The Mesolithic sites are comparatively larger. Here the hunting and gathering way of life was replaced by organized food production. Holed stones that appeared in the late Paleolithic became prominent in the Mesolithic. These holed stones appear to have been used as weights in digging sticks and as net sinkers by the fishing folks. In general, a sedentary form of living heralded the beginning of other associated cultural artifacts like pottery, living in well-built houses like wattle and clay, or of sun-dried and fired bricks.

A variety of classic collections of Microlithic tools were made. These include a thumbnail scraper to skin the small animals, with an obliquely truncated fluted core made of red corundum apart from borers and points. Usage of corundum is something unique, as it is the second hardest material known after diamond (as per the Mohs scale of hardness). The red and blue transparent varieties of corundum are the gemstones ruby and sapphire. This is the first reported usage of corundum as a microlithic tool in India, and maybe in the world.

Other tools include a Microlithic blade made of chert and quartz for cutting purposes, a Microlithic side scraper, and a Microlithic tool point with a serrated edge. Apart from these, pieces of hearth material have been collected. These were used for firing and heating, and hence are very good for dating purposes. Pieces of lightweight, hollow circlelike materials have been collected in various places. These may be slag pieces from the extraction of metal; the ancients might have had some knowledge of metallurgy.

Pottery pieces of various types have been collected. These include a broken bowl, coarse red-ware, pottery pieces embedded in mud walls, slow-wheel-turned pottery pieces, pieces of jar lid, pottery pieces with some cord impression and a very ancient fragment of pottery, possibly amongst the oldest so far collected anywhere in the world.

Even though most of the artifacts are of the Mesolithic period, there were some Paleolithic tools that were much older. It shows that people have been living in the Khambhat area for quite some time. One example is the bifacial scraper, a very characteristic Upper Paleolithic stone tool made of chert. Similar stone tools were also present—but many of them, due to long submergence and rolling in the seabed, appear to have had their edges smoothed out.

A thorough examination of the macro and micro levels of soil in the Gulf of Khambhat brought to light a wealth of plant material typical of the land domain. Extensive studies are being carried out by a botany professor, and the studies may initiate a new branch in marine archaeology, perhaps called "marine archaeobotany." So far, prominent plant species identified include palm, coconut, bamboo, and the areca plant. Many of them are fossilized.

All these were recognized under the microscope. The major wooden log, when taken out of the seabed, was very fresh, dark, and hard, and showed growth rings. Within a couple of days of exposure to the atmosphere, the wooden log shrank completely and developed cracks. It has given an important date to the area. This species has been tentatively identified as hardwood, much like rosewood. All these facts point to well-grown trees and foliage, with lots of good fresh water and a somewhat warm climate at the time of their growth in the area.

Hard dark alluvium and typical river conglomerate are found in the paleochannel below the seabed. They clearly indicate a freshwater alluvial environment. Several in situ pieces of alluvium and conglomerate have been collected. All these factors clearly establish that the paleochannels were originally well-flowing rivers in the land, which were subsequently submerged by the sea.

As some people have expressed doubts about the pottery pieces, a thorough scientific study was made involving the pottery pieces to establish their authenticity. To determine the properties of various materials, including pottery, many samples were subjected to X-ray diffraction (XRD) analysis. Since the materials that constitute pottery are clays, and a heterogeneous mixture of a variety of materials, these were analyzed accordingly. Every area has a special fingerprint pattern in the clay, which can be recognized under X-ray diffraction.

The above analysis was carried out in Deccan College, Pune, in the state of Maharashtra, India, by using an advanced instrument, the energy dispersive X-Ray fluorescence spectrometer that gave excellent results. The conclusion is that the pattern of pottery pieces corresponds very well with the locally available clay of the Gulf of Khambhat. The mineral patterns of habitational floor, wattle and daub, and land materials (alluvial deposit) are comparable. The patterns of fired clay, brick floor pieces, and vitrified clay compare very well.

All these indicate that they are genuine artifacts, made from locally available material, and are in situ. It fully confirms the presence of archaeological sites. The findings indicate that the pottery was produced locally with levigated clay, fired uniformly at about 700°C. From the presence of calcite in clay and pottery, arid to semi-arid environmental conditions prior to the submergence of the site could be deduced. Calcitized alluvial deposits indicate the existence of ancient rivers that once flowed in the submerged regions of the Gulf of Khambhat.

Dating of Samples: Most of the structures that were discovered in the Gulf of Khambhat had many similarities to the Citadel, Great Bath, and grid-iron pattern habitation site granary of the Harappan civilization. However, many of the artifacts and typology were very different and distinctive, and, with the presence of so many microtools, appeared to be much older than the Harappan. To establish the credibility and age of the civilization, it was essential to date different objects and artifacts to establish the period of the Khambhat civilization.

There are many different types of dating of archaeological artifacts—carbon dating, thermoluminescence, optically stimulated luminescence (OSL) dating, accelerator mass spectrometry (AMS) dating, dendrochronology archaeomagnetism, electron spin resonance dating, Potassium-Argon dating, and cation-ratio dating. In all, about twenty-three datable objects were selected, covering both of the paleochannels. From the samples obtained from the marine archaeological sites, it was clear that the following methods would be most suitable, and a fairly accurate age determination can be obtained from them. These are ^{14}C radiocarbon dating, radiocarbon dating by accelerator mass spectrometer (AMS), thermoluminescence (TL), and optically stimulated luminescence.

Radiocarbon dating is a method for obtaining age estimates on organic material, and is effective from the present back to a maximum of 50,000 BCE. Radioactive carbon (^{14}C) is produced in the atmosphere and is absorbed by plants; the radioactive carbon enters the human and animal cycle when the plants are eaten by animals and human beings. The absorbing of ^{14}C is stopped when a living organism dies, and ^{14}C starts to disintegrate. How much ^{14}C is disintegrated and how much is left out can be measured, and the rate at which it disintegrates is known. From this the age of organic objects like trees, corals, human remains, and shells can be determined. For age determination tests, about 50–100 g of organic material is necessary.

In the AMS, a much smaller organic sample on the order of 1–2 milligrams is enough to calculate the age. This has several advantages over the regular ^{14}C method. The thermoluminescence (TL) method is mainly used for rocks, soil materials, pottery, and other materials that were fired. It is based on the principle that almost all natural minerals are thermoluminescent. Energy absorbed from ionizing radiation frees electrons, which are trapped. Later heating releases the trapped electrons, producing light. Measurement of the intensity of the light can be used to determine how much time has passed since the last time the object was heated. Natural radioactivity causes TL to build up, so that the older an object is, the more light is produced. Since a certain amount of heating (generally up to 350°C) is required, TL works best for ceramics, cooking hearths, fired bricks, fire cracked rocks, or fire treated minerals such as flint or chert.

OSL is similar to TL dating. The minerals in the sediment grains are sensitive to light; when exposed to light, the electrons vacate the sediment grains. This process is called recombination, or a clock-setting event. To detect the age, the comparison must be made between sediment grain with a known amount of added radiation and sediment grains that are acted upon naturally. This method is suitable for a variety of unheated sediments not older than 500,000 years. This includes silty and sandy sediments that are deposited by water.

The selected samples for various types of dating were sent to some reputed institutes in India, Oxford University in England, and to Hannover, Germany. Some samples were repeated in different institutes to get confirmation of the age. The results tallied very well. The datable objects were selected to represent both of the paleochannels. However, comparatively fewer datable samples were obtained from the southern paleochannel. In the northern paleochannel, alluvial samples were collected at different depths to gain an idea about the age of alluvium, as well as the river. Of the twenty-three total samples that were dated, the ^{14}C method was followed in four samples, the TL method in six samples, and the OSL in thirteen samples.

The alluvium samples of the northern paleochannel were tested by Manipur University at the behest of NIOT. The top alluvium collected just below marine sediment was dated to around 3000 BCE, and a slightly lower alluvium gave an age of about 5000 BCE. A black alluvium that was somewhat semi-consolidated, and collected above the river conglomerate, gave an age of 19,000 BCE. Obviously the river has been flowing at least between 19,000 BCE, prior to glacial maxima, and up to 3000 BCE. This shows that the paleochannel in the north was active, and a riverine regime existed from at least about 19,000 BCE. As the area and the paleochannel to the south were proven to be a hydrocarbon-rich zone, several oil- and gas-producing wells and platforms have been put up along with Christmas trees, and several oil and gas pipelines are crisscrossing the area. Due to these factors, no further sampling beyond the preliminary one could be undertaken in view of the safety and restrictions in the oil production areas. The water depths of the alluvial samples collected in the northern paleochannel varied from 20 to 32 meters.

In the southern township or paleochannel area, six samples suitable for dating were identified. Of these, three are carbonized wooden samples, one was a sediment sample, one was a fired pottery piece, and one was hearth material. Samples from the same carbonized wood were sent to National Geophysical Research Institute, Hyderabad, India, and Geowissenschaftliche Gemeinschaftsaufgabe, Hannover, Germany, for carbon dating. This was the first sample (Location 21° 03.08' N; 72° 30.83' E) from near the southern paleochannel. This first gave a clue to the age and environment of the civilization. The calibrated age, as per NGRI, was 9580–9190 BCE and, as per Hannover Institute, was 9545–9490 BCE. It means that the age is about 9500 BCE, more than four thousand years older than the oldest city civilization of Mesopotamia and a forerunner to the Harappan civilization. But this occurred near the top of the stratigraphic column. Because of this, it was expected that at the lower levels the age would be much older, showing that the civilization could be a truly ancient one.

The wooden piece tested at Birpal Sani Institute at Lucknow,

Uttar Pradesh, gave a calibrated age of 8450 BCE. However, two important artifacts were obtained in a nearby area at lower levels. These were a nice, thin pottery and a brownish to red hearth material. A local clay sediment was also chosen along with it. All three samples were analyzed in the Physical Research Laboratory, Ahmedabad, Gujarat State, using standard thermoluminescence-based pottery dating techniques. As expected, the one of the pottery piece whose figure is given, gave a date of 13,000 ±1950 BCE. This is an important date.

Another pottery piece that was ill-fired, on OSL dating (Location 21° 12.54' N; 72° 30.370' E) by Oxford University gave an age of 16,840 ±2620 BCE. These are the oldest fired pottery pieces obtained so far in the world. Previous to this, Japanese pottery was the oldest known in the world. The "Jomon" pottery from the Fukui cave in Kyushu gave 12,000 BCE uncalibrated age. The pottery findings from Odai Yamamoto gave an uncalibrated age of 13,800–13,500 BCE. In the Gulf of Khambhat, attempts appear to have been made in experimental potterymaking. This is determined from the effects of fired clay (for making pottery), which gave the ages of 20,130 ±2170 BCE (Location 21° 13.720' N; 72° 26.190' E) and 16,600 ±1150 BCE (Location 21° 13.80' N; 72° 26.10' E) by OSL, as determined by the Oxford University dating lab. The well-fired three potteries in the northern paleochannel gave ages of 7506 ±785 BCE, 6097 ±611 BCE (both by Manipur University) and 4330 ±1330 BCE by Oxford University.

Apart from this sun-dried pottery, other pieces were collected in these areas. Three of the specimens were dated by OSL facility in Oxford. The results obtained are:

1. 31,270 ±2050 BCE
2. 25,700 ±2790 BCE
3. 24,590 ±2390 BCE

A black slipped dish, which was also sun dried, was dated in Oxford by OSL. This gave an age of 26,710 ±1950 BCE. The hearth material from the southern township (Location 21° 03.04' N; 72° 30.70' E) by TL dating from Physical Research Laboratory, Ahmedabad, gave an age of 10,000 ±1500 BCE, whereas the hearth material near the top in the northern township gave an age of 3530 ±330 BCE by OSL, at Oxford University. One of the charcoal pieces obtained on the northern side was tested by ^{14}C dating in BSIP, Lucknow. It gave the calibrated age of 3000 BCE. It tallies very well with the age of uppermost alluvium in the northern paleochannel.

The wattle and daub materials that were originally of wood and clay were seen to be burnt, but the structure of the wood was well preserved at places (being fossilized). These were tested by OSL at Oxford and by TL at Manipur from the same locations. OSL dating found it to be 5860 ±720 BCE, and TL dating determined it to be 5530 ±550 BCE. They appear to be a comparatively good match and they reflect the proper ages. They may represent the period at which these structures caught fire.

From this information it is quite clear that human activity is evident from about 31,000 BCE in what is now the Gulf of Khambhat, long before the glacial maxima at 18,000 BCE. The ancients were making potteries and getting them dried, initially in the sun. From about 20,000 BCE on, it is clear that the ancients were firing the clay to produce pottery. That means they knew how to make, maintain, and manage fire. They appear to have succeeded in making fired pottery from about 16,800 BCE. They knew the art of construction of towns and houses in neat straight lines, row after row, as picked up by side-scan sonar images (also the use of wattle and daub structures and rammed floor).

Both the northern and southern townships have continuous habitation sites interspersed with large structures in between, but good quality fired pottery makes its appearance from about 13,000

BCE on—in the southern township (we can call them metropolises), there appears to be organized activity in the form of community living, and a granary (where fossilized food grains were collected). To the south of this township, in the Gulf of Khambhat, side-scan sonar picked up a drowned dead coral colony about 1,312.3 feet (400 meters) long and 65.6 feet (20 meters) wide in water depths of about 131.2 feet (40 meters), substantiated later by sampling. It is a well-known fact that these corals live in hardly 6.6 to 9.8 feet (2–3 meters) water depth very near coastal areas. They require a clean environment and good sunlight.

Obviously, the southern metropolis appears to have been near a seacoast at a particular point of time, when the metropolis itself stood on dry land with a good free-flowing river, and was a major bustling city. The dating of coral colonies by drill core, like other places, will provide the date of the beginning of coral build-up in the area, and the top sample of coral will reveal the age at which it was drowned, giving a direct clue to the drowning of the southern metropolis. It is worthwhile to note that the datable objects, as found, only date up to 8450 BCE, based on the age of the carbonized wood.

The northern metropolis has well-made pottery pieces, wattle and daub, and so on from about 7506 BCE onward. This indicates well-organized city living. Hence it is possible that this metropolis sprang up after 8450 BCE, but long before 7506 BCE, perhaps after the submergence of the southern metropolis. The ancients appear to have shifted and founded the northern metropolis. However, the various earlier dates from sun-dried pots and other materials indicate that it was under constant habitation.

Evidence from Microlithic Tools

Apart from Paleolithic macro tools, several pieces of micro tools have been collected. Usage of such tools has been reported in

America, Europe, and other places. In South America, especially Brazil, the presence of human beings is reported from 14,000 BCE and on, coinciding with big Pleistocene mammals. Research indicated that between 17,000 and 7000 BCE, most coastal plains were lost due to sea level rise. Several Microlithic tools that were recovered are seen to occur from 10,970 BCE. But the Lagoa Santa people occupying these areas in Brazil suddenly disappeared between 8000 and 7000 BCE. In Europe, in France, Germany, Belgium, and other countries, the Microlithic tool period started around 11,800 BCE, taking back the age of the Mesolithic period. The Microlithic period spread to several areas and lasted up to the seventh millennium BCE.

In the Gulf of Khambhat, a variety of Microlithic tools have been obtained in continuation of late Paleolithic tools. The presence of highly evolved experimental pottery from 13,000 BCE, organized living, sedentary well-planned habitation, and advanced sanitary and town planning activities in the southern metropolis indicates that it had developed to be a established civilization from about 13,000 BCE. There was already evidence for the control of fire and the making of pottery from about 16,840 BCE and on. The southern metropolis has so far provided datable objects up to 8500 BCE.

The well-developed northern metropolis has dates of civilization from about 7506 BCE. One should take into consideration the ideas of Graham Hancock, who also postulated that several cultures in near-coastal areas have been flooded and submerged by rising sea levels caused by the melting of icecaps subsequent to the last ice age. The inundation maps prepared by Dr. Glen Milne of Durham University, England, clearly show that the Gulf of Khambhat area prior to 7600 BCE was mostly land, and after 6900 BCE it is mostly submerged. This type of rise in sea level is very much supported by the work of Dr. P. K. Banerjee, pertaining to the southeast coast of India, the work of Shahidul Islam and M. J. Tooley in the Bay of Bengal in Bangladesh and Sen and Banerjee's work near Calcutta.

The area is highly prone to severe seismicity. In the past five hundred years, several earthquakes have shaken the area, including the major 8.0 magnitude Richter scale event on January 26, 2001. On January 16, 1819, an 8.3 magnitude event devastated several nearby areas. These quakes cause a lot of subsidence at some places, and elevation at other places.

In the Gulf of Khambhat itself, various surveys have picked up fault zones and earthquake-affected areas with throws up to as much as 98.4 feet (30 meters) (elevation and depression). The Gulf of Khambhat was formed by a major rift. To understand the phenomenon and paleoseismic activity, NIOT commissioned Dr. Rajendran of CESS, Trivandrum, to carry out paleoseismic studies in the area surrounding the Gulf of Khambhat. His path-breaking work of identifying paleoseismic events and dating them with OSL—and also dating nearby organic material—has given excellent evidence to support the findings. He detected the presence of sand blow layers caused by old earthquakes as well as new ones. His work in peripheral land areas of the Gulf of Khambhat, like Kathana, Lotal, and Motibaur, gave evidence of major earthquakes in the Khambhat areas in the following periods:

1. 2780 ±150 BCE
2. 3983 ±150 BCE
3. 7540 ±130 BCE

Herein lies the evidence of the end of the Gulf of Khambhat civilization. In a major event at about 7600 BCE, the southern metropolis appears to have been collapsed by faulting, and sunk underwater. Because of this, the people appear to have proceeded north in the elevation higher than the sea level, and established the second or northern metropolis. This was also affected by faulting due to earthquakes around 4000 BCE, and was destroyed and submerged by an earthquake in 2780 BCE (±150 years). In this connection, it is worthwhile to point out three important aspects:

The folk songs (in local Kachchi dialogue) mention about four major towns of the ancient past. Three of these have been identified as Mohenjo Daro, Harappa, and Dholavira. (Obviously the fourth one, and the biggest and oldest of them all, is the Gulf of Khambhat metropolis.)

The second aspect is work by other agencies describing small-scale stone ruins near the Gulf of Kutch, at the present-day town of Dwarka, as the remains of the ancient, fabled city of Dwaraka—so aptly described as the abode of Lord Krishna, of *Mahabharata* fame. The city is said to have been completely transgressed by the sea, and this is vividly described by Arjuna, Krishna's main disciple in the *Mahabharata* epic.

The puzzling, incoherent aspect is the location of the city of Dwaraka. The temple we see today is hardly 900 years old. The area all around is dry, void of vegetation, and with brackish water. Krishna is supposed to have maintained a huge army of men, as well as animals such as elephants and horses. There are absolutely no trees or foliage or fresh water for a big army, so the location strikes a discordant note. But the submerged metropolis of the Gulf of Khambhat has strong, powerful flowing rivers, lots of trees and foliage, and a huge township of truly ancient times. So the metropolis in the Gulf of Khambhat could likely be the "Dwaraka City" of the *Mahabharata*.

The third aspect to consider is what happened when the first and second metropolis got submerged. It is interesting to note that there are about five hundred Harappan and pre-Harappan settlements in Gujarat, of which about 258 are on the peripheral areas of the Gulf of Khambhat. All of them are younger than the southern

Gulf of Khambhat metropolis. To the immediate west of the Gulf of Khambhat on the Saurashtra Coast is the well-known pre-Harappan and Harappan archaeological site of Padri. It has been established by the Deccan College researchers that the river Chatranji, which is now flowing east into the Gulf of Khambhat, originally flowed west, but was tilted toward the east by large-scale structural changes.

By connecting it to the southern paleochannel and extending it, it is seen to go to Prabhas Patan in the Arabian sea, a well-known pre-Harappan archaeological site mentioned in the *Mahabharata* epic. To the east, the paleochannel is seen to be an extension of the present-day Tapi River. Obviously the river was flowing right up to Prabhas Patan on the Arabian sea prior to the drifting and formation of the Gulf of Khambhat. It now falls into the Gulf of Khambhat instead of the Arabian sea. The ancients, after the catastrophe and submergence in the Gulf of Khambhat, appear to have spread out all over Gujarat and then to the surrounding areas to establish a continuing and evolving civilization of Harappan type.

So, from the foregoing it is very evident that the prehistoric civilization that matured and developed in the present-day Gulf of Khambhat was the forerunner and model of the subsequent advanced Harappan civilization known to history. This wonderful twin prehistoric metropolis of Khambhat lasted from about 13,000–3000 BCE, making it the most ancient and largest city not only in Asia but in the entire world. It is seen to be at least 7,500 years older than the oldest Mesopotamian city. However, strong evidence supports the presence of humans, from at least 31,000 BCE, who were evolving and developing, and who formed a great hitherto unknown civilization that was submerged by the flood, giving credence to local and global flood myths.

TWELVE

The Orion Zone

ANCIENT STAR CITIES OF THE AMERICAN SOUTHWEST

GARY A. DAVID

Orion of the High Desert

To watch Orion ascend from the eastern horizon and assume its dominant winter position at the meridian is a wondrous spectacle. Even more so, it is a startling epiphany to see this constellation rise out of the red dust of the high desert as a stellar configuration of Anasazi cities built between the mid-eleventh century and the end of the thirteenth century CE. The sky looks downward to find its image made manifest in the Earth, while the Earth gazes upward, reflecting on the unification of terrestrial and celestial.

Extending from the giant hand of Arizona's Black Mesa that juts down from the northeast, three great fingers of rock beckon. They are the three Hopi mesas, isolated upon this desolate but starkly beautiful landscape to which the Ancient Ones so long ago were led. Directing our attention to this "Center of the World," we clearly see the close correlation to Orion's Belt.

Mintaka, a double star and the first of the trinity to peek over the eastern horizon as the constellation rises, corresponds to Oraibi and Hotevilla on Third (or West) Mesa. The former village is the oldest continuously inhabited community on the continent, founded in the early twelfth century.

Approximately seven miles to the east, located at the base of Second (or Middle) Mesa, the village of Shungopavi (initially known as Masipa, a cognate of the Hopi underworld deity Masau'u) was reputedly the first to be established after the Bear Clan migrated into the region around the year 1100 CE. Its celestial correlative is Alnilam, the middle star of the belt.

About seven miles farther east on First (or East) Mesa, the adjacent villages of Walpi, Sichomovi, and Hano (or Tewa)— the first of which was established prior to 1300 CE—correspond to the triple star Alnitak, rising as the last of the three stars of the belt.

Nearly due north of Third Mesa's Oraibi at a distance of just over fifty-six miles is Betatakin Ruin in Tsegi Canyon, while about four miles beyond is Keet Seel Ruin. Located in Navajo National Monument, both of these spectacular cliff dwellings were built during the mid-thirteenth century. Their sidereal counterpart is the double star Rigel, the left foot or knee of Orion. (We are conceptualizing Orion as viewed from the front.)

Due south of Oraibi at an equal distance of fifty-six miles is Homol'ovi Ruins State Park, a group of four Anasazi pueblos constructed between the mid-thirteenth and early fourteenth centuries. These represent the irregularly variable star Betelgeuse, the right shoulder of Orion.

Forty-seven miles southwest of Oraibi is the primary Sinagua ruin at Wupatki National Monument, surrounded by a few smaller ruins. ("Sinagua" is the archaeological term for a group culturally similar and contemporaneous to the Anasazi.) Built in the early twelfth century, their celestial counterpart is Bellatrix, a slightly variable star forming the left shoulder of Orion.

About fifty miles northeast of First Mesa's village of Walpi is the mouth of Canyon de Chelly and its eponymous national monument. In this and its side canyon, Canyon del Muerto, a number of Anasazi ruins dating from the mid-eleventh century are found. Saiph, the triple star forming the right foot or knee of Orion, corresponds to these ruins—primarily White House, Antelope House, and Mummy Cave.

Extending northwest from Wupatki/Bellatrix, Orion's left arm holds a shield over numerous smaller ruins in Grand Canyon National Park, including Tusayan near Desert View on the south rim. Reaching southward from Homol'ovi/Betelgeuse, Orion's right arm holds a nodule club above his head. This club stretches across Mogollon Rim (an escarpment that cuts east to west across northern Arizona) and down to the Hohokam ruins near modern-day Phoenix. (The "Hohokam" was an earlier group than the two previously mentioned. They used irrigated farming methods, rather than dry.)

The head of Orion is a small stellar triangle formed by Meissa at its apex and by Phi-1 and Phi-2 Orionis at its base. It correlates to the Sinagua ruins at Walnut Canyon National Monument, together with a few smaller ruins in the immediate region.

If we conceptualize Orion not as the rectangle but as a polygon of seven sides appended to another triangle, whose base rests on the constellation's shoulders, we see that the relative proportions of the terrestrial Orion coincide with amazing accuracy. The apparent distances between the stars as we see them in the constellation (as opposed to actual light-year distances), and the distances between these major Hopi villages or Anasazi/Sinagua ruin sites, are close enough to suggest that something more than mere coincidence is at work here.

For instance, four of the sides of the polygon—A, Betatakin

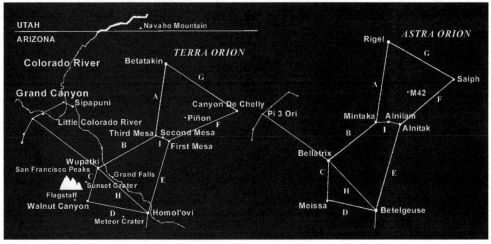

Fig. 12.1. Orion as a polygon of seven sides.

to Oraibi; B, Oraibi to Wupatki; C, Wupatki to Walnut Canyon; and F, Walpi to Canyon de Chelly—are exactly proportional, while the remaining three sides—D, Walnut Canyon to Homol'ovi; E, Homol'ovi to Walpi; and G, Canyon de Chelly back to Betatakin— are slightly stretched in relation to the constellation—from between ten miles in the case of D and E and twelve miles in the case of G. (Please see figure 12.1 above.)

This variation could be due in part either to possible carto- graphic distortions of the contemporary sky chart in relation to the geographic map, or to ancient misperceptions of the proportions of the constellation vis-à-vis the landscape. Given the physical exigen- cies for building a village, such as springs or rivers (which are not prevalent in the desert anyway), this is a striking correlation, despite these small anomalies in the overall pattern.

As John Grigsby says in his discussion of the relationship between the temples of Angkor in Cambodia and the constella- tion Draco: "If this is a fluke then it's an amazing one. . . . There is allowance for human error in the transference of the constellation on to a map, and then the transference of the fallible map on to a difficult terrain over hundreds of square kilometers with no method of checking the progress of the site from the air."*

In this case we are not dealing with Hindu/Buddhist temples, but with multiple "star cities," sometimes separated from each other by more than fifty miles. Furthermore, the "map" is actually represented on a number of stone tablets given to the Hopi at the beginning of their migrations. This geodetic configuration was influenced or even specifically determined by a divine presence—namely, Masau'u, god of death, the underworld, and the Earth plane.

Referring to the table below, we also notice the angular correspondences of Orion-on-the-Earth to Orion-in-the-Sky. Here, again, the visual reciprocity is startling enough to make us doubt that mere coincidence is responsible. Using the Bersoft Image Measurement software, however, we can correlate in degrees the precise angles of this pair of digital images seen in the diagram.

ANGLE	DEGREES	DIFFERENCE
AG Terra Orion	65.37	
AG Astra Orion	71.19	5.82
BC Terra Orion	132.6	
BC Astra Orion	130.77	1.83
CD Terra Orion	84.31	
CD Astra Orion	100.07	15.76
DE Terra Orion	97.79	
DE Astra Orion	95.65	2.14
FG Terra Orion	56.17	
FG Astra Orion	64.23	8.06

*Grigsby cited in *Heaven's Mirror: The Quest for the Lost Civilization* by Graham Hancock and Santha Faiia. New York: Crown Publishers, Inc., 1998, p. 127.

The closest correlation is between the left and right shoulders (BC and DE respectively) of the terrestrial and celestial Orions, with only about two degrees difference between the two pairs of angles. In addition, the left and right legs (AG and FG respectively) are within the limits of recognizable correspondence, with approximately 6–8 degrees of difference. The only angles that vary considerably are those that represent Orion's head (CD), with over 15 degrees of difference between terra firma and the firmament. Given the whole polygonal configuration, however, this discrepancy is not enough to rule out a generally close correspondence between Orion Above and Orion Below.

Solstice Interrelationship of Villages

Another factor that precludes mere chance in this mirroring of sky and earth is the angular positioning of the terrestrial Orion in relation to longitude. According to their cosmology, the Hopi place importance on intercardinal directions—that is, northwest, southwest, southeast, and northeast—rather than cardinal directions. The Anasazi could not, of course, make use of the compass, but relied instead upon solstice sunrise and sunset points on the horizon for orientation.

The Sun Chiefs (in Hopi, *tawa-mongwi*) still perform their observations of the eastern horizon at sunrise from the winter solstice on December 22 (azimuth 120 degrees) through the summer solstice on June 21 (azimuth 60 degrees), when the sun god Tawa is making his northward journey. On the other hand, they study the western horizon at sunset from June 21 (azimuth 300 degrees) through December 22 (azimuth 240 degrees), when Tawa travels south from the vicinity of what the Hopi call the Sipapuni to the San Francisco Peaks southwest of the Hopi mesas. (An azimuth is defined as the arc of the horizon measured in degrees clockwise from the true north point. The Sipapuni is the name for the Hopi entrance to the underworld in the Grand Canyon, located near the confluence of the Little Colorado River and the Colorado River. The San Francisco Peaks near the town of Flagstaff are the highest mountains in Arizona and are considered sacred to both the Hopi and the Navajo.)

A few days before and after each solstice, Tawa seems to stop—the term solstice literally meaning "the sun to stand still"—and rest in his winter or summer *Tawaki,* or "house." In fact, the winter Soyal ceremony is performed in part to encourage the sun to reverse his direction and return to Hopiland instead of continuing southward and eventually disappearing altogether.

The key solstice points on the horizon that we designate by the azimuthal degrees of 60, 120, 240, and 300 degrees (only at this specific latitude, however) recur in the relative positioning of the Anasazi sky cities. For instance, if we stand on the edge of Third Mesa in the village of Oraibi on the winter solstice, we watch the sun set at exactly 240 degrees on the horizon, directly in line with the ruins of Wupatki almost fifty miles away. The sun disappears over Humphreys Peak, the highest mountain in Arizona, where the major shrine of the *katsinam* (also spelled *kachinas,* beneficent supernatural beings who act as spiritual messengers) is located.

Incidentally, if this line between Oraibi and the San Francisco Peaks is extended southwest, it intersects the small pueblo called King's Ruin in Big Chino Valley, once a stop-off point on the major trade route from the Colorado River. If the line is extended still farther southwest, it intersects the mouth of Bill Williams River on the Colorado.

Inhabited from 1026 CE (or possibly earlier in light of the underlying pit house) to about 1300 CE, King's Ruin has a thirteen-room foundation, twelve of which could have been two stories high. The 500 pieces of unworked shell found at the site indicate substantial trade with the Pacific. Necklaces of turquoise, black shale, and

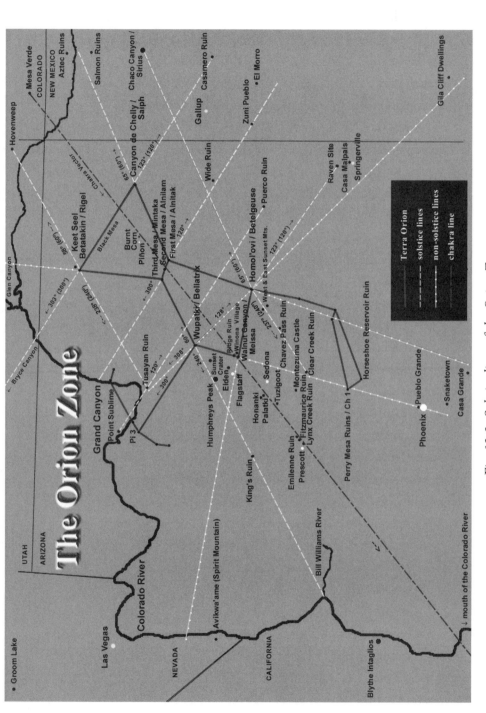

Fig. 12.2. Solstice lines of the Orion Zone.

red argillite were also found. One argillite necklace, consisting of 2,031 beads, stretched sixty-six inches long. Fifty-five graves were also discovered, containing sixty-six individuals, most of which were buried in the extended posture with heads oriented toward the east, awaiting Pahana's (or the White Brother's) return.

If we stand at Wupatki during the summer solstice, we see the sun rise directly over Oraibi on Third Mesa at 60 degrees azimuth on the horizon. On that same day the sun sets at 300 degrees azimuth, to which the left arm of the terrestrial Orion points. In addition, from Oraibi the summer solstice sun sets 12 degrees north of the Sipapuni near the Colorado River, the "Place of Emergence" of the Hopi from the previous Third World (or era) to the current Fourth World.

If we perch on the edge of Canyon de Chelly, not looking downward into the canyon but instead southwest at the winter solstice sunset, the sun on the horizon appears about 5 degrees south of the First Mesa village of Walpi. This line, extended farther southwest beyond the horizon, intersects both Sunset Crater, a volcano that erupted in 1064 CE, and Humphreys Peak, elevation 12,633 feet.

Again, the reciprocal angular relationship between the two pueblo sites remains, so from Walpi during the summer solstice the sun appears to rise from Canyon de Chelly fifty miles away. A northeastward extension of this 65 degree line eventually reaches a point in New Mexico near Salmon Ruin and Aztec Ruin. Salmon Ruin near the San Juan River was occupied for a few generations after 1088 CE, then abandoned and reoccupied between 1225 CE and the late 1200s.

The pueblo contained between 600 and 750 rooms. It also had a tower kiva built on a platform twenty feet high that was made of rock imported from thirty miles away. Ten miles north of Salmon is Aztec Ruin (an obvious misnomer), located on the Animas River. During its peak development it contained about 500 rooms. Like the former, this latter site was originally inhabited in the early twelfth century by people of Chaco Canyon, and then re-inhabited from 1225 to 1300 CE by people of Mesa Verde. In addition, it has a restored Great Kiva.

A winter solstice sunrise line (120 degrees azimuth) drawn from Walpi past Wide Ruin traverses the Zuni Pueblo (the Zuni are a tribe closely related to the Hopi) and ends just south of El Morro National Monument. Inhabited from 1226 to 1276 CE, Wide Ruin (or Kin-Tiel) is located about fifty miles due south of Canyon de Chelly. It is an oval-shaped pueblo of 150–200 rooms with a number of kivas. Atsinna pueblo, located atop a high mesa at El Morro National Monument, was a mid-thirteenth century rectangular structure, part of which was three stories in height. It had 500 to 1,000 rooms and two kivas, one circular and the other square.

If during the winter solstice we stand at the rim of Tsegi Canyon (wherein the ruins of Betatakin and Keet Seel are located) and sight southeast along the edge of Black Mesa, we see the sun come up over Canyon de Chelly and Canyon del Muerto. The sun is in fact at 120 degrees azimuth on the horizon directly over Antelope House Ruin in the latter canyon. An extension of the same line into New Mexico intersects Casamero Ruin.

Constructed in the mid-eleventh century, Casamero Ruin was a small thirty-room pueblo. However, its Great Kiva, one of the largest in the Southwest, was seventy feet in diameter—even slightly more spacious than the better-known Casa Rinconada at Chaco Canyon about forty-five miles to the north.

From the same spot at Tsegi Canyon, also on the first day of winter, we see the sun set at 240 degrees azimuth over Grand Canyon more than eighty miles to the southwest. From Tsegi, a summer solstice sunrise line of 60 degrees azimuth additionally intersects Hovenweep National Monument in southeastern Utah, well known for the archaeo-astronomical precision of its solstice and equinox markers. Again from Tsegi, a sunset line of 300 degrees azimuth crosses Bryce Canyon National Park and Paunsaugunt Plateau in southern Utah, where nearly 150 small Anasazi and Fremont ruins have been identified.

If we travel 112 miles almost due south of Tsegi Canyon to the ruins at Homol'ovi and look fifty miles to the northwest, the summer solstice sunset appears 8 degrees south of Wupatki. This line (designated as "H" in the table on page 17) running between Homol'ovi and Wupatki passes between Grand Falls (an impressive cataract along the Little Colorado River) and Roden Crater (a volcanic cinder cone that artist James Turrell has turned into an immense earth sculpture) to finally end at Tusayan Ruin on the south rim of the Grand Canyon.

Again, from the reciprocal village of Wupatki, the winter solstice sun rises 8 degrees north of Homol'ovi. This Wupatki-Homol'ovi line (128 degrees azimuth) extending southeast passes just south of Casa Malpais Ruin and ends less than ten miles south of Gila Cliff Dwellings in New Mexico. Located in the Mogollon Mountains of west-central New Mexico, Gila Cliff Dwellings National Monument is a ruin comprised of forty rooms in five separate caves located 150 feet above the canyon floor. The timbers of these structures have been tree-ring dated to the 1280s. The late Mogollon (or Mimbres) people are known for their exquisite black-on-white pottery, using realistic though stylized designs. The site was abandoned by 1400 CE.

From Homol'ovi, a winter solstice sunrise line (120 degrees azimuth) falls 7 degrees north of Casa Malpais and 3 degrees north of Raven Site Ruin, both north of the modern town of Springerville, Arizona. Casa Malpais is a thirteenth century Mogollon site of a hundred rooms with a square Great Kiva (one of the largest in the Southwest). It also has catacombs, ceremonial rooms, three winding stone stairways, and an astronomical observatory. Because of the nature of the artifacts found, such as crystals, ceremonial pipes, and soapstone fetish stands, it is thought to have been primarily a religious center.

Located nearly ninety miles southeast of Homol'ovi and about twelve miles north of the Casa Malpais, the Raven Site (privately owned by the White Mountain Archaeological Center) was occupied as early as 800–1450 CE, and had more than eight hundred rooms and two kivas.

From Homol'ovi during the winter solstice sundown (240 degrees azimuth), the sun passes directly through East and West Sunset Mountains, the gateway to the Mogollon Rim. This line from Homol'ovi proceeds past the early fourteenth century, thousand-room Chavez Pass Ruin on Anderson Mesa (known in Hopi as Nuvakwewtaqa, "mesa wearing a snow belt") and continues along the Palatkwapi Trail down to Verde Valley, ending near Clear Creek Ruin.

If the summer solstice sunrise line (60 degrees azimuth) is extended from Homol'ovi into New Mexico, it intersects the vicinity of Chaco Canyon, the largest and perhaps most impressive of all the Anasazi sites in the Southwest. In the overall astral-terrestrial pattern Chaco corresponds to Sirius, the brightest star in the heavens located in the constellation Canis Major.

In this schema, each village is connected to at least one other by a solstice sunrise or sunset point on the horizon. The interrelationship provided a psychological link between one's own village and the people in one's "sister" village miles away. Moreover, it reinforced that the divinely ordered coordinates of the various sky cities come down to earth. Not only had Masau'u/Orion spoken in a geodetic language that connected the Above with the Below, but Tawa/ Sun had also verified this configuration by his solar measurements along the curving rim of the *tutskwa,* or sacred earth.

Non-Solstice Lines,
the Grand Chakra System, and the
Hopi Winter Solstice Ceremony

In addition to the solstice alignments, a number of intriguing non-solstice lines exist to corroborate the pattern as a whole. As heretofore stated, an extension of the solstice line between Oraibi and Wupatki (the belt and left shoulder of the terrestrial Orion respectively) would ultimately end on the Colorado River at the point where a major trail east toward Anasazi territory began. Similarly, if the non-solstice line between Walpi and Homol'ovi (the belt and the right shoulder respectively) is extended, it intersects the wrist of the constellation and terminates within five miles of the important Hohokam ruin site and astronomical observatory of Casa Grande Ruins National Monument, near the Gila River 150 miles to the south.

We have also already discussed the extension of the Walpi-Canyon de Chelly solstice line (Orion's right leg), ending up at the Salmon-Aztec ruins area. An extension of the Oraibi-Betatakin non-solstice line (Orion's left leg) would bring us to Glen Canyon National Recreation Area, nearly fifty miles to the north. Ruefully, hundreds or perhaps even thousands of small Anasazi ruins were submerged by the construction of the Glen Canyon Dam in 1963, and the few that remain can only be reached by boat.

Another alignment of ancient pueblo sites forms the grand chakra system of Orion and indicates the direction that his spiritual energy flows. Drawing a line southwest from Shungopavi/Alnilam on Second Mesa, we pass less than five miles southeast of Roden Crater and Grand Falls, both mentioned above. Continuing southwest, the line runs by Ridge Ruin, through Winona Village, and into the forehead of Orion—namely, Walnut Canyon National Monument, a significant mid-twelfth century Sinagua ruin located in the foothills of the San Francisco Peaks.

Occupied from 1085–1207 CE, Ridge Ruin was a thirty-room pueblo with three kivas and a Maya-style ball court. It was also the site of the so-called Magician's Burial. Thought by Hopi elders to be of the Motswimi, or Warrior Society, this apparently important man was interred with twenty-five whole pottery vessels and over 600 other artifacts, including shell and stone jewelry, turquoise mosaics, woven baskets, wooden wands, arrow points, and a bead cap. Winona Village, which was occupied at the end of the eleventh century, contained about twenty pit houses and five surface storage rooms.

If the line is extended farther still, it intersects the red rock country of Sedona with its electromagnetic vortexes, passing the small but gorgeously located ruin and pictograph (rock painting) site of Palatki (or "Red House"), as well as the larger Honanki (or "Bear House"). In Verde Valley the newly energized vector directly transits Tuzigoot National Monument, a major thirteenth century Sinagua ruin of over a hundred rooms perched on a hilltop for the probable purpose of stellar observation.

This line traverses the Black Hills of Arizona and goes by the newly excavated Emilienne Ruin in Lonesome Valley, which had a foundation of twelve rooms, most of which could have been two stories high, plus eleven outlying one-room units. The line then intersects the Fitzmaurice Ruin, located upon a ridge on the south bank of Lynx Creek in Prescott Valley. The Fitzmaurice Ruin, occupied from 1140 to 1300 CE, had twenty-seven rooms in which were found beads, pendants, bracelets, and eighty-one amulets, including crystals, animal fetishes, obsidian nodules (so-called Apache Tears), and a curious six-faceted, truncated pyramid carved from jadeite and measuring one-and-a-half centimeters wide.

The line continues through the small Lynx Creek Ruin at the northern base of the Bradshaw Mountains, treks across the northern limits of the Sonoran desert, and passes near the geoglyphs on the Arizona/California border to ultimately reach a point just north of the Colorado River's mouth. Similar to the Nazca lines of Peru, these geoglyphs (also called intaglios) of human, animal, and star figures, some over a hundred feet long, were made by removing the darker, "desert-varnished" pebbles to expose the lighter soil beneath, as relayed by authors Jefferson Reid and Stephanie Whittlesey in *Archaeology of Ancient Arizona.*

According to the Mohave and Quechan tribes of the lower Colorado River region, the human figures represent the deity Mastamho, the creator of the Earth and all life. Notice the similarity between the name of this god and that of the Hopi earth god Masau'u. These figures are thought to be between 450 and 2,000 years old.

The same line extended in the other direction from Shungopavi travels northeast across Black Mesa, passing just southeast of Four Corners to finally end up at the major Anasazi sites at Mesa Verde National Park in southwestern Colorado.

In this series of villages, we see a total of eleven both major and minor Anasazi or Sinagua ruins and one Hopi pueblo perfectly aligned over a distance of more than 275 miles within the framework of the tellurian Orion. The probability that these were randomly distributed is highly unlikely and increases the possibility that Masau'u (or some other divine agent) directed their positioning.

This "ley line" forms a grand chakra system that provides an inseparable link and a conduit of pranic earth energy flowing from the Hopi mesas to the evergreen forests of the San Francisco Peaks. More specifically, Walnut Canyon symbolizes the terrestrial Orion's third eye, or pineal gland, which is etymologically derived from the Latin word *pinus,* or "pine cone."

At this point we might ask: Why was the template of Orion placed on the Earth at the specific angle that we find it relative to longitude? The chakra line mentioned above, which runs in part from Shungopavi/Alnilam (the belt of Orion) to Walnut Canyon/ Meissa (the head of Orion), is 231 degrees azimuth in relation to Shungopavi. The direction of southwest is 225 degrees azimuth. The axis for the terrestrial Orion is thus within 6 degrees of northeast/ southwest.

Imagine we are standing at Shungopavi shortly after midnight on the winter solstice nine centuries ago—about the time when the Hopi mesas were first settled. Looking southwest, we find the middle star of Orion's Belt hovering directly above the *southwestern* horizon at an altitude of approximately 38 degrees. Specifically, at 1:15 a.m. on December 22, 1100 CE, Alnilam was at 231 degrees azimuth. In other words, as we gaze from the central star of the earthbound belt of Orion toward his head located in the foothills of the San Francisco Peaks where the sacred *katsinam* live, we see the celestial constellation precisely mirror the angle of the terrestrial configuration.

Also at 1:15 a.m. on this date, Bellatrix is at 240 degrees azimuth, and Meissa is at 242 degrees azimuth. Forty minutes later, Alnilam is at 240 degrees, the azimuthal degree at which the sun will set at 5:15 p.m. on this same day. Incidentally, at the winter solstice sunset time, Orion is just rising on the opposite horizon, thus emphasizing the pivotal relationship of Orion/Masau'u and the Sun/Tawa. All astronomical computations were performed with Skyglobe (made by the now-defunct KlassM Software).

But what is the significance of this precise time when the middle star in Orion's Belt is at 231 degrees? At the very moment we are watching this sidereal spectacle, "one of the most sacred ceremonies" of the Hopi, known as the Soyal, is also taking place in the kiva, or subterranean prayer chamber. Just past his meridian, Orion can be clearly seen through the hatchway overhead. This is the time "when Hotomkam [Orion's Belt] begins to hang down in the sky."

Now a powerful, barefooted figure descends the kiva ladder.

White dots that resemble stars have been painted on his arms, legs, chest, and back. He carries a crook on which has been tied an ear of black corn—Masau'u's maize that signifies the Above. One account identifies him as Muy'ingwa, the germination deity related to Masau'u. Another calls him "Star Man," ostensibly because his headdress, made of four white corn leaves, represents a four-pointed star; perhaps Aldebaran in the Hyades.

This ritually transmuted person takes a hoop covered with buckskin and begins to dance. His "sun shield," fringed with red horsehair, is about a foot in diameter with a dozen or so eagle feathers tied to its circumference. Its lower hemisphere is painted blue, its upper right quadrant is red, and its upper left quadrant is yellow. Two horizontal black lines for the eyes and a small downward pointing triangle for the mouth are painted on the lower half of this striking face representing Tawa.

Alexander Stephen, who witnessed the rite at Walpi in 1891, remarked that the star priest stamps upon the *sipapu* (or a hole in the floor of the kiva that links it to the underworld) as a signal to start the most important portion of the ceremony. *This occurs just after 1:00 a.m.,* the time on this date in 1100 CE when Orion was at *231 degrees* azimuth.

As the dance rhythm crescendos, the "Star Man" begins to twirl the sun hoop very fast in clockwise rotation around the intercardinal points between two lines of singers—one at the north and the other at the south. By his "mad oscillations" (to use Alexander Stephen's phrase), the star priest is attempting to turn back the sun from its southward journey. "All these dances, songs, and spinning of the sun are timed by the changing positions of the three stars, Hotomkam, overhead. Now is the time this must be done, before the sun rises and takes up his journey."

If this were merely a solar ritual, we assume that it would take place at sunrise. On the contrary, the sidereal position of Orion must reflect the terrestrial positioning of the constellation, which occurs only after the former has passed its meridian, that is, "when Hotomkam begins to hang down in the sky." Prior to dawn, runners are sent out to the shrines of *both* Masau'u (Orion) and Tawa (the sun) to deposit *pahos* (or prayer feathers), offerings to the two gods whose complex interaction helps to assure the seasons' cyclic return, keeping the world in balance for yet another year.

Egyptian Parallels to the Arizona Orion

In their bestseller *The Orion Mystery,* Robert Bauval and Adrian Gilbert have propounded what is known as the Orion Correlation Theory. The coauthors have discovered an ancient "unified ground plan" in which the pyramids at Giza form the pattern of Orion's Belt. According to their entire configuration described just briefly here, the Great Pyramid (Khufu) represents Alnitak, the middle pyramid (Khafre) represents Alnilam, and the slightly offset smaller pyramid (Menkaure) represents Mintaka.

In addition, two ruined pyramids—one at Abu Rawash to the north and another at Zawiyat Al Aryan to the south—correlate to Saiph and Bellatrix respectively. Farther south, three pyramids at Abusir correspond to the triangular head of Orion. Bauval and Gilbert also believe that the pyramids at Dahshur—namely, the Red Pyramid and the Bent Pyramid—represent the Hyades stars of Aldebaran and Epsilon Taurus, respectively.

This schema furthermore correlates Letopolis, located due west across the Nile from Heliopolis, with Sirius, the most brilliant star in the sky. As coauthor Gilbert states in *Signs in the Sky:*

> It was Bauval's contention that the part of the Milky Way which interested the Egyptians most was the region that runs from

the star Sirius along the constellation of Orion on up toward Taurus. This region of the sky seemed to correspond, in the Egyptian mind at least, to the area of the Memphite necropolis, that is to say the span of Old Kingdom burial grounds stretching along the west bank of the Nile from Dashur to Giza and down to Abu Ruwash. At the centre of this area was Giza; this, he determined, was the earthly equivalent of Rostau (Mead's Rusta), the gateway to the Duat or underworld.

The region in Hopi cosmology corresponding to the Duat is called Tuwanasavi (literally, "Center of the World"), located at the three Hopi mesas. Similar to the ground-sky dualism of the three primary structures at the Giza necropolis, these natural "pyramids" closely reflect the Orion's Belt stars. In addition, the entry to the nether realms is known in Hopi as the Sipapuni, located in the Grand Canyon. This culturally sacrosanct area mirrors the left arm of Orion. Whereas the Egyptian Rostau is coextensive with the axis mundi of the Belt stars formed by the triad of pyramids, the Hopi gateway to the underworld in Grand Canyon is adjacent to the centerplace but still close enough to be archetypally resonant.

In a later book entitled *The Message of the Sphinx,* Robert Bauval and coauthor Graham Hancock describe the cosmic journey of the Horus-King, or son of the sun, to the underworld: "He is now at the Gateway to Rostau and about to enter the Fifth Division [Hour] of the Duat—the holy of holies of the Osirian afterworld Kingdom. Moreover, he is presented with a choice of 'two ways' or 'roads' to reach Rostau: one which is on 'land' and the other in 'water.'"

The discussion by Bauval, Gilbert, and Hancock of the Egyptian master plan is a great deal more complex than what is merely sketched out in this article. Their opus involves the precession of the equinoxes, star-targeted shafts in the Great Pyramid, and other topics that are not directly relevant to our discussion. However, this compelling work has challenged many orthodox ideas in Egyptology and spawned heated debates at both the amateur and professional levels.

We have been blessed with a wealth of Egyptian hieroglyphic texts, both on stone and on papyrus, from which we can reconstruct the Egyptian cosmology. Unless we consider petroglyphs more as a form of linguistic communication than as rock "art," then the Hopi and their ancestors, on the other hand, had no written language; hence we must rely on their recently transcribed oral tradition. In this regard the *tawa-mongwi* (or "sun watcher") Don Talayesva from Oraibi Village describes an interesting parallel to Rostau.

As a young man attending the Sherman School for Indians in Riverside, California, during the early years of the twentieth century, Talayesva became deathly ill and, in true shamanistic fashion, made a journey to the spirit world. After a long ordeal with many bizarre, hallucinatory visions, he reached the top of a high mesa and paused to look. The following is his account from his book *Sun Chief: An Autobiography of a Hopi Indian.*

> Before me were two trails passing westward through the gap of the mountains. On the right was the rough narrow path, with the cactus and the coiled snakes, and filled with miserable Two-Hearts making very slow and painful progress. On the left was the fine, smooth highway with no person in sight, since everyone had sped along so swiftly. I took it, passed many ruins and deserted houses, reached the mountain, entered a narrow valley, and crossed through the gap to the other side. Soon I came to a great canyon where my journey seemed to end; and I stood there on the rim wondering what to do. Peering deep into the canyon, I saw something shiny winding its way like a silver thread on the bottom; and I thought that it must be the Little Colorado

River. On the walls across the canyon were the houses of our ancestors with smoke rising from the chimneys and people sitting out on the roofs.

In this narrative, the dry, narrow road filled with cacti and rattlesnakes, where progress is measured at just one step per year, is contrasted with the easy, broad road quickly leading to the canyon of the Little Colorado River. A few miles east of the confluence of this river and the Colorado River is the previously mentioned Sipapuni, the actual location of the Hopi "Place of Emergence" from the Third World to the Fourth World. Physically, it is a large travertine dome in the Grand Canyon to which annual pilgrimages are made to gather ritualistic salt.

In correlative terms, the Milky Way is conceptualized as the "watery road" of the Colorado River at the bottom of the Grand Canyon—that sacred source to which spirits of the dead return to exist in a universe parallel to the pueblo world they once knew. (In an alternate conception, this stellar highway traverses the evergreen forests of the San Francisco Peaks, upon whose summit a mythical kiva leading to the underworld is located.)

Talayesva's account also includes such traditionally otherworldly motifs as the "Judgment Seat" on Mount Beautiful, which supports a great red stairway, at least in his vision. (This peak is actually located about eight miles west of Oraibi.) We furthermore hear of a confrontation with the Lord of Death, in this case a threatening version of Masau'u (the Hopi equivalent of Osiris), who chases after him.

Like the Egyptian journey to the Duat, the Hopi journey to Maski (literally, "House of Death") has two roads—one on land and the other on water. In this context we must decide if the latter is really a code word for the sky. In the "double-speak" of the astral-terrestrial correlation theory, are these spirits in actuality ascending to the celestial river of the Milky Way? Is this, then, the purpose of the grand Orion schema? To draw a map on the Earth that points the way to the stars?

Returning to the subject of Orion projected upon the deserts of both Egypt and Arizona, we find both discrepancies and parallels. In terms of distinction, the Egyptian plan is on a much smaller scale than the one incorporating the Arizona stellar cities, using tens rather than hundreds of miles. Furthermore, in the Egyptian schema the bright stars of Betelgeuse and Rigel are perplexingly unaccounted for.

The Giza terrestrial Orion from head to foot is oriented southeast to northwest, while the Arizona Orion is oriented southwest to northeast. Of course, the pyramids are located west of the Nile River, while the Hopi mesas are located east of the "Nile of Arizona," namely, the Colorado River.

We should also point out that Abusir is not in the correct location to match Orion's head on the constellatory template. Bauval and Gilbert state that Abusir is "a kilometer or so southeast of Zawiyat al Aryan," which is Bellatrix (Orion's left shoulder). It is, in fact, about 3.7 miles (6 kilometers) southeast. In other words, Abusir is nearly four miles south-southeast of where it should be according to the Orion Correlation Theory. Unlike Bauval, Gilbert, and Hancock, I have not been to Egypt, but the consultation of any scale map will verify this statement.

Despite these few differences, the basic orientation of the Egyptian Orion is similar to that of the Arizona Orion—that is, south, the reverse of the celestial Orion. According to Dr. E. C. Krupp of the Griffith Observatory, this is one factor that invalidates the Orion Correlation Theory. His critique, however, is the result of a specific cultural bias in which an observer is looking down upon a map with north at the top and the south at the bottom.

Imagine, instead, that we are standing on top of the Great

Pyramid—or, for that matter, at the southern tip of First Mesa in Arizona—and gazing southward just after midnight on the winter solstice. The other two pyramids, or mesas, would be stretching off to the southwest in a pattern that reflects the belt of Orion now achieving culmination in the southern sky. We can further imagine that if the upper portion of the terrestrial Orion were simply lifted perpendicular to the apparent plane of the earth while his feet were still planted in the same position—Abu Rawash and an undetermined site in the case of Egypt; Canyon de Chelly and Betatakin in the case of Arizona—then this positioning would perfectly mirror Orion as we see it in the sky.

When the Anasazi gazed into the heavens, they were not looking at an extension of the physical world the way we perceive it today. They were instead witnessing a manifestation of the spirit world. Much like the Egyptian Duat, the Hopi underworld encompasses both the vast reaches of the sky and the region beneath the surface of the earth. This fact is validated by the dichotomous existence of ancestor spirits who live in the subterranean realm, but who periodically return to their earthly villages as clouds bringing the blessing of rain.

Is it simply just another coincidence that the Hopi word *tu'at,* also spelled *tuu'awta,* meaning "hallucination" or "mystical vision," sounds so close to the Egyptian *Duat*—spelled by E. A. Wallis Budge, former director of antiquities at the British Museum, as *Tuat,* that seemingly illusory realm of the afterlife?

Even though the eastern and western domains ruled by Tawa (the sun) remain constant, the polar directions of north and south—controlled by the Elder and Younger Warrior Twins (sons of the sun), respectively—are reversed. Therefore, the right hand holding the nodule club is in the east and the left hand holding the shield is in the west, similar to the star chart. The head is, however, pointing roughly southward instead of northward. This inversion is completely consistent with Hopi cosmology, as the terrestrial configuration is conceptualized as a reversal of the spirit world, of which the sky is merely another dimension. (An alternate explanation for the change in directions is the possibility that the pole shift, which destroyed the Hopi's Second World, reversed the position of the constellation's mundane aspect.)

Gazing up at Orion on a midwinter night, we can imagine that *our* perspectives have switched. Suspended high above the land, we stare southwest toward the sacred *katsina* peaks and the head of the celestial Masau'u suffused in the evergreen forests of the Milky Way. Ironically, it is here on the high desert of Arizona that we also intuit the truth of the hermetic maxim attributed to the Egyptian god Thoth (later known as Hermes Trismegistus): "As above, so below."

Let's briefly look at one more so-called coincidence: The word *zona,* as in the name Arizona, is particularly significant. From classical times onward, it has literally been defined as "The girdle [or Belt] of Orion."

THIRTEEN

On the Possibility of Instantaneous Shifts of the Poles

HAS IT OCCURRED DURING HUMAN EXISTENCE?

FLAVIO BARBIERO

It is well known that the poles have often changed their position on the Earth's surface during past geological eras. The marks left by thick ice sheets in Africa and India, the residual magnetism in ancient rocks, the old coral reefs' and coal deposits' distribution and so on are compelling evidence that the poles have wandered from what is today's equator to the actual poles. According to geological data, this wandering should have happened in "jumps."

Scientists attribute it to the drift of continents and to the displacement of large quantities of materials, due to erosion and sedimentation processes, which in theory can provoke a very slow shift of the poles: a few centimeters per year at the most, but in hundreds of millions of years it can result in shifts of thousands of miles.

There is strong evidence, however, that this phenomenon could be much faster.

In his book *The Path of the Pole,* Charles Hapgood expresses the hypothesis that the poles have changed their position three times during the past 100,000 years. Between 50,000 and 12,000 years ago, at the end of Pleistocene, the North Pole was located somewhere around Hudson Bay in eastern Canada, and it moved to its current position in the twelfth millennium BCE.

They were eccentric with respect to the actual poles, suggesting a different position for them. To support his thesis, Hapgood presents an impressive quantity of evidence, which can be summarized as follows:

1. Between 50,000 and 12,000 years ago an impressive ice cap, more than two miles thick, spread from the Hudson Bay area southward, down to the latitude of New York, and westward to join, at its maximum extent, glaciers flowing down from the Rocky Mountains in Alaska. During the same period, northern Europe was covered by ice caps, which at their maximum extent reached the latitude of London and Berlin. The quantity of water trapped in these ice sheets and in the glaciers scattered around the world was so large that the sea level was about 426.5 feet (130 meters) lower than today.

2. The current "scientific" explanation for the existence of these ice caps is that they were created by a cooler climate all over the world. But this theory is contradicted by the absence, during the ice age, of ice sheets in Siberia, which was actually populated, up to its northernmost regions, by one of the most impressive zoological communities of all time. Millions (more than 40 million, according to F. C. Hibben) of mammoths roamed Siberia and Alaska. Animals as large as this can be found today only in tropical regions, or in other areas where the supply of fodder is guaranteed year round.

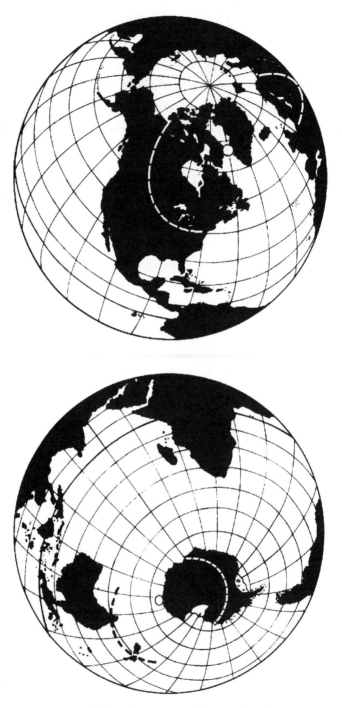

Fig. 13.1. An extension of the polar ice caps during the Pleistocene era.

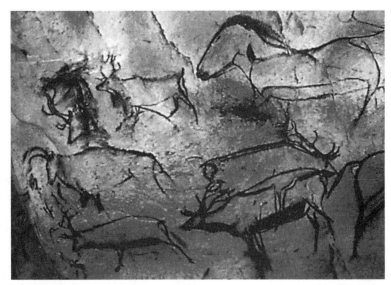

Figs. 13.2. The Chauvet cave in southern France contains hundreds of animal paintings dating from Paleolithic times (ca. 31,000 BCE).

Fig. 13.3. The representation of humans in cave paintings was rare. More often than not, animals such as the reindeer, depicted here, were featured prominently.

Fig. 13.4. The mammoth illustrated in this cave painting from Chauvet cave in France apparently coexisted with a large number of many other types of fauna and mega fauna, both artic and tropical, leaving scholars puzzled about the climate of the time.

Fig. 13.5. In the cave paintings of Chauvet cave, predators such as the lion are often depicted alongside such animals as the rhino.

Fig. 13.6. The hyena and the leopard shared the same environment as camels, buffalo, giant beavers, and many other types of animals that existed during the Pleistocene era.

Fig. 13.7. Horses appear in the imagery of the the Chauvet cave, a series of chambers and vestibules all covered with masterworks breathtaking in their use of the sophisticated techniques of perspective and shading.

It is counterintuitive that during the ice age one of the largest zoological communities since the dinosaurs existed in those very areas that are today regarded, due to their extreme climatic conditions, as among the most hostile on Earth. With the mammoths there were dozens of other animal species, the majority of which are extinct today. Of these species we have a great number of skeletons, complete animals that have been preserved in the permafrost, and many wonderful paintings in Paleolithic caves. The oldest among the latter is Chauvet Cave in France, parts of which were painted as early as 33,000 years ago. It contains paintings of breathtaking beauty.

The unknown artists have, with a few strokes, represented to perfection animals that at the time were living in the plains of central Europe (and at the same time in Siberia and Alaska)—but the beauty of the paintings makes the zoologist wonder in more ways than one. How could such a varied assembly of animals coexist? To what bizarre ecological environment could such motley fauna belong? We find the reindeer next to the rhinoceros; the mammoth, with its woolly mantle, near the hippopotamus; lions side by side with bears, leopards, and Przewalski's horses. There are also giant beavers and sloths, big-horned deer, camels, sabertooth tigers, buffaloes, aurochs, and many other species.

These images represent an incredible mixture that leaves us puzzled and astonished. Arctic and tropical fauna together, on the same plain, in perfect balance with the environment! Such an extraordinarily varied and numerous animal community—the likes of which can be found nowhere on Earth today—seems to challenge current opinion on climatic conditions during the ice age. Moreover, this community suddenly disappeared when the ice age ended—exactly at the moment when, according to modern theories, climatic conditions were supposed to have become milder and more supportive of life.

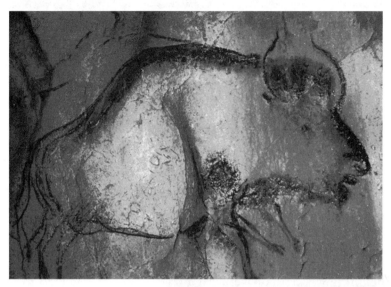

Fig. 13.8. Animals that symbolized strength, such as the buffalo rendered here, are featured frequently on the cave's walls.

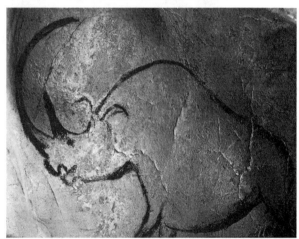

Fig. 13.9. The prehistoric artworks that are displayed in the cave are so sophisticated that specialists in ice age art initially assumed that the works were created in recent times.

Fig. 13.10. The ancient ibex was a wild goat with large, recurved horns. It was apparently revered by the ancient peoples who painted its image on the walls of the cave.

This mysterious, now-vanished fauna populated the Siberian islands well inside the Arctic Sea; their remnants can be found on islands located at only 621.4 miles (1,000 kilometers) from the North Pole, and rock engravings have been found in the same islands. This strongly suggests that in the late Pleistocene (a period during which the global climate was supposedly *much colder* than it is today, especially at the high latitudes), the Arctic Sea was in fact *much warmer* than it is today.

3. On the other side of the world, the climate was cooler in Australia and New Zealand, back then partially covered by large glaciers—but there is solid evidence that Antarctica, now completely covered by a thick layer of ice, must have been partially free of it at that time. Sediment cores, collected in the Weddell area, show that in the late Pleistocene large rivers must have flowed in this part of Antarctica. Again, is this not a strong suggestion that the climate of part of Antarctica must have been much milder during the ice age than it is even today—despite the intense global warming experienced in the past century?

A shift of the poles, occurring around 11,500 years ago, could explain completely and coherently the climatic situation before that date, and the situation that came into being after that date.

The hypothesis that the position of the poles—as well as the inclination of the terrestrial axis in relation to the ecliptic—might change first started to gain serious consideration during the nineteenth century. Some of the greatest scientists of the time, including J. C. Maxwell and Sir George Darwin (son of the famous Charles Darwin), considered this problem and decided that the stabilizing effect of the equatorial bulge was so great that no conceivable force could make the Earth shift on its axis except for a collision with another planet. They therefore dismissed the idea of any shift of the poles as impossible and, in fact, not worth discussing. Their influence has been so strong that to this day no one has seriously contemplated such a hypothesis again.

Hapgood also uncritically accepts the assumption that only a "planetary collision" is capable of displacing the axis of rotation. Therefore he proposes a theory that explains the shift of the poles as the result of the shift of the whole Earth's crust. Based on the research of the Russian scientist V. V. Beloussov, he assumes that at a depth of approximately one hundred miles in the upper mantle there is a layer of liquid rock, which behaves as a bearing allowing the whole crust to "shift" when subjected to a displacing force. In Hapgood's opinion, this force is provided by the centrifugal momentum of ice caps positioned eccentrically with respect to the poles. In this way the Earth would keep its axis of rotation unchanged, but the poles and the whole Earth's surface would shift and change latitude.

The evidence proving that the poles were in different positions during the Pleistocene era is quite impressive, and this explains why Hapgood's theory was approved by scientists of the caliber of Einstein and K. F. Mather. Yet it meets with so many difficulties that it appears highly controversial. Above all, it does not seem to be compatible with other geological theories that are widely accepted today, notably plate tectonics and continental drift.

Furthermore, the theory does not explain some of the key peculiarities of the climate changes of the late Pleistocene—most significantly the speed with which these changes appear to have taken place. According to Hapgood's theory, it took the North Pole at least two thousand years to move from its previous position to the present. The evidence we have, however, supports a much faster climatic change. It was Hapgood himself who underlined a great amount of data proving the high speed at which the shift of the poles appears to have happened; yet the mechanism he proposes does not explain this speed.

It appears that we can completely and coherently explain what took place at the end of the Pleistocene by admitting the possibility of a shift of the poles of the same magnitude Hapgood hypothesizes, but in a much shorter time: not more than a few days. This possibility is generally rejected because no convincing explanation for such a phenomenon has so far been brought forward. According to scientists, the only way to make a planet change its axis of rotation is by "adding" to it a mass comparable to its own. However, it is my view that at least one other way exists, one that has not yet been considered—that of "reshaping" its equatorial bulges around a different axis.

If Earth was a perfectly rigid and spherical body, a single man walking on its surface could make its poles shift. In fact, the stability of the Earth is provided only by its equatorial bulges, some 7.4 miles (12 kilometers) thick—very small with respect to the Earth's radius. Move the equatorial bulges and the poles will move accordingly. Impossible? Not really, if you consider that two thirds of Earth's surface is covered with water; every naval engineer knows that free liquid surfaces induce instability (experience shows that a simple tsunami is enough to make the poles shift by a measurable amount). Let's see how this instability can result in a wide and permanent shift of the poles.

The Apollo Objects

Severe consequences can follow when the Earth is hit by a comet or an asteroid. We know for certain that this happened many times in the past and can happen again at any moment in the future.

If the Earth didn't have oceans and atmosphere, its surface would be marked with huge numbers of craters, like the moon and Mercury. On our planet, instead, erosion and sedimentary processes very quickly erase the traces of collisions with asteroids, meteorites, and comets. Only where recent ice sheets have scraped the surface, thus uncovering the traces of ancient collisions, as in parts of

Canada, is it possible to count the craters accurately. Based on this approach, G. W. Wetherill has estimated that in the last 600 million years our planet has been hit by at least 1,500 objects with a diameter larger than one kilometer each.

The majority of these collisions are caused by a class of celestial bodies that astronomers have named "Apollo objects," a class of asteroids whose perihelion lies inside the orbit of the Earth. The first of these objects was discovered by Reinmuth on 1932 and named Apollo, which gave the name to the class. At present, more than one hundred Apollos of a diameter of at least one kilometer are known. The largest discovered so far, Hephaistos, has a diameter of 6.2 miles (10 kilometers). The total number of Apollo objects with a diameter of half a mile (one kilometer) or more is estimated to be between one and two thousand.

As the perihelion of the Apollos lies inside the orbit of the Earth, it follows that they periodically have the chance to collide with it. The probability of such an event is estimated at 5×10^{-9} per year per single Apollo. Therefore we have a probability of at least four collisions each million-year period with objects as large as one kilometer or more. As the size of the objects becomes smaller, this probability grows exponentially to reach one impact every few centuries for objects of 328.1 to 656.2 feet (100–200 meters) diameter.

The direct effects of a collision with an Apollo-like object are devastating. Author Tom Gehrels estimates that a half mile (one kilometer) wide object, colliding with the Earth at a speed of 12.4 miles (20 kilometer) per second, would liberate an energy equivalent to ten billion Hiroshima-type nuclear bombs.

How the Poles Can Shift

Scientists are trying to understand what the overall effect on the Earth of a collision with an Apollo object could be. The scenarios they have come up with so far look rather dire. After all, many believe that the extinction of the dinosaurs followed an impact with an asteroid. None of these scenarios, however, has taken into consideration the possibility that such an impact could also provoke an almost instantaneous shift of the poles. This is because, compared to the Earth, a one-kilometer-wide asteroid is like a tiny sphere of two millimeters next to a ball of 82 feet (25 meters). Its mass is absolutely negligible. The displacement of the poles directly due to it, if any, can be measured in the order of centimeters.

What can't be neglected, however, is the *torque* provoked by the impact. Due to the very high speed of the asteroid, the impulsive torque it delivers can be of sufficient magnitude to overcome, for an instant, the reaction torque developed by the Earth. The torque, in itself, lasts for too short a period to produce any measurable effect; yet I will argue that it can trigger a process that in the end results in a change of the axis of rotation.

Let's see how.

Earth is a gyro. A gyro subject to a disturbing force reacts with a movement called "precession." The precession phenomenon has been studied exhaustively—but, unfortunately, only for the case when the precession's rotation is much smaller than the gyro's main rotation, the only interesting case for technical applications. Scientists, therefore, are not familiar with the case in which the two rotational components have the same order of magnitude. This case is examined here in the appendix, where the behavior of a gyro subject to a disturbing torque of increasing value is shown. It appears that when the torque reaches a critical value, equal to the maximum reaction torque that can be developed by the gyro, the latter changes its axis of rotation permanently. The new axis, which coincides with the previous precession's axis, is maintained even if the disturbing torque diminishes again, as long as its value is higher than zero. Only if and when the torque is completely null (or becomes negative), does the gyro recover its previous rotational axis.

The behavior of the Earth when subjected to a disturbing torque is obviously the same. In fact, the Earth has a movement of precession due to the disturbing torque exercised by the sun-moon gravitational attraction on the equatorial bulges. This torque is one million times smaller than the maximum reaction torque that can be developed by Earth. Simple calculations, however, allow us to establish that an object as small as a half-kilometer-wide asteroid, hitting the planet in the right spot and at the right angle, is capable of developing an impulsive torque of the same magnitude of the maximum Earth's reaction torque. In this case the Earth assumes, for a very short instant, a different axis of rotation.

If at the moment of the impact the force of the sun-moon gravitational attraction on the equatorial bulge has the same direction as the force developed by the impact, a shift of the poles will inevitably follow. Immediately after the impact, in fact, the torque should go down to zero, and the Earth should recover its previous rotational axis. But if the torque exerted by the sun-moon attraction has the same direction, the torque cannot be zeroed and therefore the Earth keeps "memory" of the impact and of its direction. This "memory" consists of an extremely small rotational component, with the same direction as that of the impact, on the order of one millionth of the normal rotation.

What is particular in this rotational component is that it is fixed with respect to the Earth. If the latter were a solid gyroscope, this situation would stay unchanged indefinitely. The planet, however, is not homogenous and rigid. First of all, it is covered by a thin layer of water, which reacts immediately to any change of motion. Second, even the "solid" outer shell is in reality plastic and can be easily "reshaped" by centrifugal forces.

Under the effect of this tiny rotational component, seawater begins to move toward a circle perpendicular to that rotation (the new equator). This is a very small effect, and if it were the only component, the resulting equatorial bulge would be of a few feet only. But as this happens, the value of the rotational component increases at the expense of the main rotation, therefore increasing the centrifugal force that makes more water move toward the new equator, thus increasing the force and so on. This process starts very slowly, but accelerates progressively, until the centrifugal force developed by this rotational component grows strong enough to induce deformations of the Earth's mantle.

From here on, the equatorial bulge is quickly "reshaped" around the new axis of rotation and Earth will soon be stable again, with a different axis of rotation and different poles.

This mechanism shows that the Earth's poles, contrary to what has always been postulated, can make "jumps" in a matter of days (that is, almost instantaneously) of thousands of miles, due to the effects of forces at first sight negligible, such as the impact of a medium-size asteroid and the sun-moon gravitational attraction on the equatorial bulge, combined with the effects of water mobility and the plasticity of the crust.

Phenomena That Occur during a Shift of the Poles

Suppose that the Earth has been hit by an asteroid and that the conditions to trigger a shift of the poles have been met. On the basis of the adjustments necessary to reshape the equatorial bulge around the new axis of rotation, and the consequent reestablishment of the isostatic equilibrium of the crust, we can predict what kind of phenomena would happen on the surface.

Some areas of the Earth's crust would be driven to move upward, others downward. The up and down movements necessary to reshape the bulge would be different from site to site. For a shift of 20 degrees, for example, the movements would be about 2,200 to 2,600 feet (700–800 meters) at the most. Very small, compared to the diameter of the Earth, but nonetheless of great consequence on the surface. We know that the mechanisms that maintain the isostatic equilibrium of the mantle are very effective; so there is no doubt that after a while the equilibrium would be reestablished around the new axis of rotation, with poles and equator in different positions.

It is important to evaluate how long it would take for this to happen. We know that the layers of the crust, when subjected to a force over a certain limit, break suddenly, causing an earthquake. In the situation we have hypothesized, only seawater would be displaced at the beginning, with a gradual increase of the speed of rotation around the new axis. When the rotational speed reaches a certain critical value, sudden adjustments of the mantle would begin to happen, and from that moment on the process would be sharply accelerated, and the reshaping of the bulge would be completed in a very short time.

How short? Weeks, days, or hours? Impossible to say. A simulation with a mathematical model should give reliable results. The process of reshaping the equatorial bulge should follow a course of exponential type: after the initial sharp peak, it should decrease very quickly. Adjustment phenomena, however, are expected to continue for a long time, as the isostatic equilibrium is reestablished more and more accurately.

Obviously, readjustments of the equatorial bulge of that size cannot happen without causing extensive fractures of the crust, which would provoke earthquakes of tremendous magnitude predictably stronger than the most devastating known today. A sudden burst of volcanic activity in all areas subjected to strain is also expected.

The beginning of adjustments of the crust would start more than just earthquakes and volcanic activity. On the whole the oceans' water and the atmosphere follow the rotational movement of the Earth, but they are not tied to it—if the Earth should suddenly change the direction of its rotation, thanks to their inertia, they would keep up their previous motion at first, thus provoking a dreadful hurricane all over the continents, with violent winds and torrential rains. The continents would be swept by hurricane force winds, reaching speeds of hundreds of miles per hour. However, the attrition with the Earth's surface would soon force them to follow the new movement.

The water of the oceans would play a much greater destructive role. We must expect wide fluctuation of sea level in many parts of the world, predictably in the order of hundreds of yards.

The same reasoning applies where the center of Earth is concerned, consisting of a solid core, surrounded by a liquid layer of iron. This would at first continue in its motion, naturally undergoing strong attrition in the mantle boundary region that would create turbulence, which might have important effects. According to the latest theories, the liquid iron layer is the site of electrical currents, which are responsible for the Earth's magnetic field. This turbulence could provoke perturbation in the magnetic field that might even lead to an inversion of the magnetic poles.

An important element to evaluate the climatic conditions following a shift of the poles is the inclination (tilt) that the new axis of rotation will assume with respect to the ecliptic. This has a tremendous effect on the climate. According to the mechanism we have spelled out so far, the axis of rotation that the Earth would acquire at the moment of impact should depend on the direction of the hit. However, it is impossible to predict what the actual direction of the new axis would be once stabilized. It is certain that it would not be the same as the previous one, unless by fortuitous chance. Therefore, following a shift of the poles the course of the seasons would very likely be different.

For example, on the hypothesis that the axis of rotation is almost vertical with respect to the ecliptic, there would be an enormous growth of ice at high latitudes and altitudes, with subsequent lowering of sea level. On the other hand, the climate would be much more stable than it is today, with very limited (or nonexistent) seasonal climatic differences and an uninterrupted growth of vegetation. This would bring about the disruption of today's climatic barriers, with subsequent spreading of tropical species toward northern regions and vice-versa. There would also be the maximum possible development of ecological communities.

This appears to be the situation that existed in the Pleistocene era, when imposing zoological communities thrived at the very edge of the ice caps and the phenomena that put an end to this situation appear to be the same as we have described.

The tenth millennium BCE appears to be critical under several aspects. It was precisely in that period that the Paleolithic cultures, which had thrived for more then thirty millennia, all of a sudden disappeared. And all over the world, both in land and at sea, most of a megafauna that thrived for more than 100,000 years declined to extinction.

The Pleistocene geological era came to an end, marked by a burst of volcanism and terrible earthquakes, witnessed by the collapse of the vaults in many European and American caves, and a dreadful hurricane all over the world. Even the magnetic field underwent serious challenges and nearly inverted—not to mention the climate, which from then on underwent a radical change.

Twelve thousand years have elapsed since then, a period that is only twice the length of the historical period. We are the direct descendants of men who managed to survive those cataclysms. Is it possible that the memory of events of that kind has been completely canceled in such a relatively short time? Myths and legends about an apocalyptic disaster, which marked the "beginning" of humankind—the "universal flood"—are common to almost all populations of the world. Some of those legends, starting from Plato's, report not only the same kind of phenomena we have described, but even the same date.

Is this only a mere coincidence, or are they referring to the phenomena that put an end to the Pleistocene era only 12,000 years ago? If we consider that a tiny asteroid can provoke an almost instantaneous shift of the poles, the second alternative definitely looks the most likely.

FOURTEEN

Mysterious Strangers

NEW FINDINGS ABOUT THE
FIRST AMERICANS

GRAHAM HANCOCK

I have consistently argued that the Americas were inhabited in prehistoric times by a variety of different ethnic groups—Negroid, Caucasoid, and Mongoloid. Such ideas have caused deep offense to some American Indians, who have long claimed to be the only "native" Americans, and also contradict academic teachings that hold that the New World remained uninhabited by humans until the end of the last ice age, and was then settled exclusively by Mongoloid nomads from Asia who crossed the Bering Strait about 12,000 years ago (when there was a land bridge between Siberia and Alaska) and made their way thence into all of North and South America—reaching the latter only about 9,000 years ago. Naturally, this teaching also holds that no Caucasoids or Negroids were present anywhere in the Americas prior to the coming of Columbus and the European conquest in the fifteenth and sixteenth centuries CE.

One of the historical mysteries that drew me into writing

Fig. 14.1. An image of Viracocha. Provided courtesy of Santha Faiia at www.GrahamHancock.com.

Fig. 14.2. Caucasoid figure known as the "Walker" or La Venta. Provided courtesy of Santha Faiia at www.GrahamHancock.com.

Fingerprints was the eloquent mythical and sculptural testimony concerning a time, long ago, when people who were definitely not American Indians inhabited the Americas. Both the god Viracocha, in South America, and the god Quetzalcoatl in Mexico were described as tall, white-skinned, and red-bearded—sometimes blue-eyed as well.

At Monte Alban, near Oaxaca, and at La Venta, on the Gulf of Mexico (a site associated with the mysterious "Olmec" culture, supposedly the first and the oldest high civilization of Mesoamerica),

Fig. 14.3. Caucasoid figure, Monte Alban. Provided courtesy of Santha Faiia at www.GrahamHancock.com.

ancient carved figures have been found that seem to depict such individuals. In the case of Monte Alban, these Caucasoid figures date back to about 600 BCE and, in the case of La Venta to about 1200 BCE—almost three thousand years before the European conquest.

More intriguing still, other sculptures, mostly in the form of carved megalithic heads, were also found at La Venta in the same archaeological strata as the Caucasoid figures. Once again, these sculptures, the so-called Olmec Heads, do not display the

Fig. 14.4. Olmec Head, Monte Alban. Provided courtesy of Santha Faiia at www.GrahamHancock.com.

typical features of Native American Indians. This time they are unmistakably Negroid in appearance—depicting individuals who must have closely resembled modern Africans, Melanesians, or Australian Aborigines.

In *Fingerprints of the Gods* and *Heaven's Mirror,* I wrote at length about these anomalistic sculptures and the myths that accompany them. I argued that real people must have served as the models for both types and that they should therefore be taken seriously as historical testimony of the presence of Caucasoids and Negroids in the New World more than three thousand years ago.

Fig. 14.5. Olmec Head. Provided courtesy of Santha Faiia at
www.GrahamHancock.com.

Fig. 14.6. Olmec Head. Provided courtesy of Santha Faiia at www.GrahamHancock.com.

This view was not accepted by a single orthodox scholar in 1995 when *Fingerprints* was first published. Since then, new evidence has come to light, which has obliged the experts to reconsider their position and step back from the dogma of exclusively Mongoloid settlement of the Americas. The first breakthroughs came in 1996 and 1997, as published in the *Washingon Post,* Final Edition, Tuesday, April 15, 1997.

Skeletons unearthed in several western states and as far east as Minnesota are challenging traditional views that the earliest Americans all resembled today's Asians. The skeletons' skulls bear features similar to those of Europeans, suggesting that Caucasoid people were among the earliest humans to migrate into the New World more than 9,000 years ago. Anthropologists have known of such bones for years, but did not fully appreciate their significance until re-appraising them over the last few months. The new analyses were prompted by the discovery last summer of the newest addition to the body of evidence—the unusually complete skeleton of an apparently Caucasoid man who died about 9,300 years ago near what is now Kennewick, Washington. . . . The man's head and shoulders were mummified, preserving much of the skin in that area. . . . Those who examined [him at first thought that the bones] were the remains of a European settler [until radiocarbon revealed their great age]. "It's an exciting time, and I think we're going to see some real changes in the story about the peopling of North America," said Dennis Stanford, an authority at the Smithsonian Institution's National Museum of Natural History.

Not all scholars agree that Kennewick Man was a Caucasoid. But at the very least the discovery has raised significant doubts about the established model of the peopling of the Americas. Other discoveries have raised further doubts.

Similarities to Modern-Day Aborigines or Africans

In *Fingerprints of the Gods,* I described one of the colossal Olmec Heads of La Venta:

It was the head of an old man with a broad flat nose and thick lips. The lips were slightly parted, exposing strong, square teeth.

The expression on the face suggested an ancient, patient wisdom, and the eyes seemed to gaze unafraid into eternity. . . . It would be impossible, I thought, for a sculptor to invent all the different combined characteristics of an authentic racial type. The portrayal of an authentic combination of racial characteristics therefore strongly implied that a human model had been used.

I walked around the great head a couple of times. It was 22 feet in circumference, weighed 19.8 tons, stood almost 8 feet high, had been carved out of solid basalt, and clearly displayed an authentic combination of racial characteristics. Indeed, like other pieces I had seen, it unmistakably and unambiguously showed a Negro. . . . My own view is that the Olmec Heads present us with physiologically accurate images of real individuals of Negroid stock. . . .

I returned to this mystery in my book *Heaven's Mirror:*

Orthodox historians do not accept the presence of any Africans in the New World prior to the time of Columbus and have tried to sidestep the implications of the obviously African features of the three-thousand-year-old Olmec Heads—sixteen of which have so far been found. It may at least be taken as a sign that there is no racism in archaeology that there are also supposed to have been no Caucasians in the New World before Columbus! Scholars have therefore predictably raised quibbles about the Quetzalcoatl myth of the tall bearded white man and have sought to dismiss any suggestions that it might be reflected in the numerous reliefs of Caucasian faces that have been excavated in some of the oldest archaeological sites of Mexico. In the Olmec area, several were found in the same strata as the African heads and sometimes side by side with them, but images of Caucasians have also been excavated as far afield as Monte Alban in the south-west, a site dated to between 1000 and 600 BCE.

In 1996 and 1997 the discovery of Caucasian bones more than nine thousand years old in the Americas seems, quite suddenly, to have validated the Quetzalcoatl myth. It is therefore now legitimate to ask how long it may be before another lucky turn of the archaeologist's spade will uncover the bones of individuals who could have served as prototypes for the great Olmec Heads.

That "lucky turn of the archaeologist's spade" was not long in coming. On August 22, 1999, the *London Sunday Times* (and, a few days later, a BBC2 television documentary) reported the discovery in Brazil and Columbia of more than fifty skeletons and skulls of a Negroid people who had lived in South America about 12,000 years ago—about 3,000 years before the first known penetration of Mongoloid peoples into this region.

One particularly well-preserved example, the remains of a young girl whom scientists have nicknamed "Luzia," has been described as "the oldest human skeleton yet found on the American landmass." It has been studied by Walter Neves, professor of biological anthropology at the University of Sao Paolo, who states:

> When we started seeing the results, it was amazing because we realized the statistics were not showing these people to be Mongoloid; they were showing that they were anything except Mongoloid. . . . They are similar to modern-day Aborigines or Africans and show no similarities at all with Mongoloids from east Asia and modern-day Indians.

The *Sunday Times* also quotes Richard Neave, a forensic artist with the University of Manchester, who has made a reconstruction of Luzia's face. "That," he comments, "is a Negroid face.

The proportions of the face do not say anything about it being Mongoloid."

Up till now, as far as I know, no scholar has pointed out that the discoveries in Brazil and Columbia could offer an explanation for the Negroid features of the Olmec Heads. It's true that the Olmec sculptures were found in strata between three and four thousand years old—whereas the Brazilian and Columbian skeletons are much older than that. But this does not rule out a possible connection. At the very least, it is surely an intriguing coincidence (a) that pieces of monumental sculpture depicting Negroid individuals have come down to us from the prehistoric antiquity of the Americas and (b) that a Negroid people, previously unidentified and unsuspected by historians, have now been scientifically proven to have inhabited the Americas around 12,000 years ago. Perhaps the "Olmec" Heads were not made by the "Olmecs" at all, but were inherited by them as heirlooms, handed down from an earlier time?

A Chinese Influence?

A footnote to this story, and a sign of the galloping collapse of consensus amongst orthodox scholars concerning the peopling of the Americas, is a row that began to simmer in academic circles in the late 1990s concerning possible cultural links between ancient Chinese and ancient American cultures—specifically the Olmec and the Shang. The main proponents of this view are Professor Mike Xu, who teaches in the foreign languages department at the University of Central Oklahoma in the United States, and Chen Hanping of China's Historical Research Institute.

According to an article published in *U.S. News and World Report,* and to Internet postings, Xu believes that "the first complex culture in Mesoamerica may have come into existence with the help of a group of Chinese who fled across the seas as refugees at the end of the Shang dynasty. The Olmec civilization arose around 1200 BCE, which coincides with the time when King Wu of Zhou attacked and defeated King Zhou, the last Shang ruler, bringing his dynasty to a close."

Xu is also reported to have "explosive" evidence in the form of archaic writings:

> Over the past three years he has found some 150 glyphs on photographs of real specimens of Olmec pottery, jade artifacts, and sculptures. As well as himself leafing through dictionaries of ancient Chinese, he has also taken his drawings of these markings to be examined by mainland Chinese experts in ancient writing, and most have agreed that they closely resemble the characters used in Chinese oracle bone writings and bronze inscriptions. "At first these experts all tried to send me away, saying they could not give an opinion of foreign artifacts," Mike Xu recalls. But after his repeated entreaties, they reluctantly took a look. The moment they saw his drawings, each of them asked: "Where in China were these inscriptions found?"

When they heard they came from America, they were all dumbstruck. "If these inscriptions had been found in excavations in China," says Chen Hanping, a research associate at the mainland's Historical Research Institute, "they would certainly be regarded as symbols of the pre-Quin-dynasty period."

Reaction from other scholars has been almost universally hostile. This posting, from Constance Cook, associate professor of Chinese at Lehigh University, sums up several key objections:

> Some asked me to post my observations re: the script on the Olmec Celts identified by Chen, Hanping as Chinese in *U.S. News and World Report*. I have finally seen the article with the

reproduction of the Olmec graphs and the section that Chen believed was similar to the oracle bone script of the Shang.

1. The graphs isolated by Chen are not Chinese. They bear some graphic similarity to some archaic Chinese graphs or parts of graphs but as single graphs equal nothing and do not have the equivalents he assigned to them. It is bogus.

2. Obviously, the graphs/glyphs pulled out by Chen should be considered within the context of the entire "inscription." This is impossible as the rest of the marks bear none but a few isolated similarities. In fact the Olmec "script" may not represent language at all, but like the Naxi and other ur-scripts, be more a code for storytelling than an actual transcription of language. The Shang oracle bone script, on the other hand, is very advanced and unquestionably qualifies as belonging to a writing system.

3. Finally, the "inscription" must be considered within the context of the sculptures. There is very little beyond the occasional face of human representation in Shang period art (some carved jade figures, but these are kneeling, often incised, and covered with animal décor, tattoos, and clothes). One famous bronze has a shaman-like figure in the mouth of an animal, but there is no similarity to the Olmec representations.

4. A point of correction: The *U.S. News and World Report* article claims that Chen is the foremost authority of only about twelve scholars worldwide who are trained in ancient script. First, Chen is a very minor scholar. Second, there are more than twelve scholars in the United States alone who can read Shang script, many many more in China and elsewhere.

Bulging Eyes and a Long, Curling Nose

I haven't the faintest idea whether Professor Cook and the other critics are right or whether Xu and Chen are onto something with their Shang-Olmec connection. Further open-minded research is required to settle the issue, but in the meantime I would like to

Fig. 14.7. Sanxingdui mask. Provided courtesy of Santha Faiia at www.GrahamHancock.com.

Fig. 14.8. Mayan Chac mask, architectural feature. Provided courtesy of Santha Faiia at www.GrahamHancock.com.

draw attention to a small piece of evidence that Santha and I first became aware of when we visited the "Mysteries of Ancient China" exhibition at the British Museum in early January 1997 and, more recently, when we visited Beijing's History Museum in March 1999.

We were intrigued by artifacts from the sacrificial pits at Sanxingdui in Sichuan province, including bronzes in the shape of human heads, fragments of gold, many jades, and a large number of elephant tusks; these artifacts were discovered in 1986 and are the work of a previously unidentified high civilization that flourished in Sichuan from around 1200 to 1000 BCE.

Of particular interest was an exhibit described as "Mask with ornamented forehead and protruding pupils." It is one of three large masks that were found in Pit Two at Sanxingdui. According to the British Museum's commentary, found in *Mysteries of Ancient China:*

> The most startling features are the pupils of the eyes, which project on stalks. . . . A further remarkable feature is the long upstanding projection rising from the nose of the mask. This projection is scroll-shaped, with an upright section coiled at the top and with a double loop at the bottom. . . . The combination of the large ears, the protruding eyes and the tall quill makes this face completely fantastic.

Is it a coincidence that almost exactly the same "fantastic" and "startling" features—bulging eyes and a long, curling projection to the nose—are found upon the Chac masks of the ancient Maya of Central America, the successors to the Olmecs? Chac masks were sometimes also incorporated as architectural features into Maya temples.

Figs. 14.9–10. Step pyramids, Xian, China. Provided courtesy of Santha Faiia at www.GrahamHancock.com.

Conspicuous Strangers

Perhaps such similarities are just coincidences—although personally I rather doubt it. Setting aside all other issues and anomalies, however (and there are many), I contend that the ancient Caucasoid and Negroid skeletons that have been found in the New World mean that there can no longer be any room for preconceived notions. The true history of the peopling of the Americas is likely to turn out to be extraordinarily complex and attenuated, involving many different ethnic groups and cultures in many different epochs. It does not surprise me at all that the Chinese might have been here three thousand years ago, or the Phoenicians at about the same time—as others have suggested—or the remarkable Jomon culture of Japan, or the Egyptians, or—much later—the Vikings. I believe it to be very likely, indeed little short of a certainty, that all these peoples and several others as well must have independently "discovered" the Americas, in isolated individual cases, over and over again, from the very earliest times.

But how early?

If we keep on pushing this inquiry back into the past—back beyond the earliest-known historical civilizations—then what, ultimately, do we come to?

The possible Shang influence on Olmec writing and art takes us back 3,000 years, well within the bounds of recorded history. But Kennewick Man and Luzia take us back, respectively, 9,300 and 12,000 years, the latter date being some 7,000 years earlier than Sumer and Egypt—the earliest known historical civilizations—and smack in the middle of the end of the last ice age, when the Earth passed through a gigantic cataclysm and floods from the melting ice sheets scoured the globe.

Some scholars have expressed the view that Luzia's people could have come to South America from the Pacific and been related to Australian Aborigines. Certainly there is evidence that in ancient times the Aborigines possessed sophisticated sailing and navigational skills. So far, however, there is no historical theory that can explain the presence in the New World at such early dates not only of Negroid Aborigines but also of Caucasoids—let alone the fact that memories of these conspicuous strangers appear to have been preserved in myths and in sculptures as far afield as Bolivia and Mexico.

As well as being an intriguing human and archaeological mystery, therefore, what we now know about the first Americans represents a radical and robust challenge to orthodox scholarship.

I suspect that this story will run and run.

Selected Bibliography

Chapter 1. Thoughts on Parapsychology and Paranormal Phenomena

Carroll, Sean. "Telekinesis and Quantum Field Theory." Go to http://blogs.discovermagazine.com/cosmic variance and search on "Telekinesis and Quantum Field Theory."

Gurney, Edmund, Frederic W. H. Myers, and Frank Podmore. *Phantasms of the Living.* London: Rooms of the Society for Psychical Research, Trübner and Co., 1886.

Lynch, Gary, and Richard Granger. *Big Brain: The Origins and Future of Human Intelligence.* New York: Palgrave Macmillan, 2008.

Schoch, Robert M., with Robert Aquinas McNally. *Pyramid Quest: Secrets of the Great Pyramid and the Dawn of Civilization.* New York: Tarcher/Penguin, 2005.

———. *Voices of the Rocks: A Scientist Looks at Catastrophes and Ancient Civilizations.* New York: Harmony Books, 1999.

———. *Voyages of the Pyramid Builders: The True Origins of the Pyramids from Lost Egypt to Ancient America.* New York: Tarcher/Penguin, 2003.

Schoch, Robert M., and Logan Yonavjak, eds. *The Parapsychology*

Revolution: A Concise Anthology of Paranormal and Psychical Research. New York: Tarcher/Penguin, 2008.

Tyson, Neil deGrasse. "The Beginning of Science." *Natural History,* March 2001.

Yukteswar Giri, Jnanavatar Swami Sri. *The Holy Science.* Los Angeles: Self-Realization Fellowship, 1974 [New Edition; Introduction dated 1894].

www.cosmicvariance.com/2008/02/18/telekinesis-and-quantum -field-theory (Accessed 7 March 2008.)

Chapter 2. Quantum Philosophy and the Ancient Mystery School

Capra, Fritjof. *The Tao of Physics.* Halifax, NS: Shambhala, 2000.

De Lubicz, Schwaller. *The Temple of Man.* Rochester, Vt.: Inner Traditions, 1981.

Mercer, Samuel, trans. *The Pyramid Texts.* Charleston, S.C.: BiblioLife, 2009.

Neumann, Erich. *The Origin and History of Consciousness.* Princeton, N.J.: Princeton University Press, 1995.

Zukav, Gary. *The Dancing Wu Li Masters: An Overview of the New Physics.* New York: HarperOne, 2001.

Chapter 3. The Egypt Code

Bauval, Robert. *The Egypt Code.* New York: The Disinformation Company, 2008.

———. *The Orion Mystery: Unlocking the Secrets of the Pyramids.* New York: Three Rivers Press, 1995.

BBC2. "Great Pyramid: Gateway to the Stars." 1995.

Malek, Dr. Joromir. *Discussions in Egyptology,* 1994.

Chapter 4. Alternative History and Esoteric Philosophy

BBC2. "Atlantis Uncovered." *Horizon,* October 28, 1999.

———. "Atlantis Reborn." *Horizon,* November 4, 1999.

———. "Atlantis Reborn Again." *Horizon,* December 14, 2000.

Bleibtreu, John. *The Parable of the Beast.* England: Paladin, 1976.

Booth, Mark. *The Secret History of the World.* New York: Overlook, 2008.

Redford, Donald, ed. "The History of Egyptology." In *The Oxford Encyclopedia of Ancient Egypt.* Oxford: Oxford University Press, 2000.

West, John Anthony. *Serpent in the Sky.* Wheaton, Ill.: Quest, 1993.

www.insideoutthinking.co.uk.

Chapter 5. An Open Letter to the Editors of *Archaeology*

Fagan, Garrett. *Bathing in Public in the Roman World.* Ann Arbor: University of Michigan Press, 1999.

NBC. "The Mystery of the Sphinx." November 10, 1993.

West, John Anthony. "Atlantis and Beyond: The Lure of Bogus Archaeology." *Archaeology,* May/June, 2003.

Chapter 6. Dark Mission

Brookings Research Institute. "Proposed Studies on the Implications of Peaceful Space Activities for Human Affairs." Washington, DC, December 14, 1960.

FOX. "Conspiracy Theory: Did We Land on the Moon?" February 15, 2001.

Hoagland, Richard C., and Mike Bara. *Dark Mission: The Secret History of NASA*. Port Townsend, Wash.: Feral House, 2007.
www.apolloarchive.com/apollo_gallery.html.
www.history.nasa.gov/alsj/frame.html.
www.hq.nasa.gov/office/pao/History/ap15fj/index.html.
www.spacearchive.info.

Chapter 7. History and Celestial Time

Cruttenden, Walter. *The Great Year*. The Yuga Project, 2003.
———. *Lost Star of Myth and Time*. Pittsburgh, Pa.: St. Lynn's Press, 2005.
De Santillana, Giorgio, and Hertha von Dechend. *Hamlet's Mill: An Essay Investigating the Origins of Human Knowledge and Its Transmission through Myth*. Jaffrey, N.H.: Godine Press, 1977.
Diamond, Jared. *Guns, Germs, and Steel: The Fates of Human Societies*. New York: W. W. Norton, 2005.
Yogananda, Paramahansa. *Autobiography of a Yogi*. Los Angeles: Self-Realization Fellowship, 2006.
Yukteswar, Sri. *The Holy Science*. Los Angeles: Self-Realization Fellowship, 1990.

Chapter 8. The Orion Key

Bauval, Robert. *The Orion Mystery: Unlocking the Secrets of the Pyramids*. New York: Three Rivers Press, 1995.
Bauval, Robert, and Graham Hancock. *Keeper of Genesis/Message of the Sphinx*. Portsmouth, N.H.: William Heinemann Ltd., 1996.

Chapter 9. The Cygnus Mystery

Collins, Andrew. *The Cygnus Mystery*. London: Watkins Books, 2006.

Sagan, Carl. *Carl Sagan's Cosmic Connection: An Extraterrestrial Perspective*. Cambridge: Cambridge University Press, 2000.

Sagan, Carl, and I. S. Shklovskii. *Intelligent Life in the Universe*. New York: Dell, 1966.

www.sondela.co.uk

Chapter 10. Electromagnetism and the Ancients

Barrow, Lennox. *Irish Round Towers*. London: Jarrold and Sons, 1976.

Burke, John, and Kaj Halberg. *Seed of Knowledge, Stone of Plenty: Understanding the Lost Technology of the Ancient Megalith-Builders*. Tulsa, Okla.: Council Oak Books, 2005.

Callahan, Philip S. *Ancient Mysteries Modern Visions: The Magnetic Life of Agriculture*. Austin, Tex.: Acres USA, 2001.

———. "The Mysterious Round Towers of Ireland: Low Energy Radio in Nature." *The Explorer's Journal,* Summer 1993.

Ellul, Joseph S. *Malta's Prediluvian Culture at the Stone Age Temples*. Malta: Printwell Ltd., 1988.

Hancock, Graham. *Underworld: The Mysterious Origins of Civilization*. New York: Crown, 2002.

Imbrogno, Philip, and Marianne Horrigan. *Celtic Mysteries in New England*. Woodbury, Minn.: Llewellyn, 2000.

Chapter 11. The Gulf of Khambhat

Buck, William. *Mahabharata*. Berkeley: University of California Press, 2000.

Chapter 12. The Orion Zone

Allen, Richard Hinckley. *Star Names: Their Lore and Meaning.* New York: Dover, 1963.

Barnett, Franklin. *Excavation of Main Pueblo at Fitzmaurice Ruin: Prescott Culture in Yavapai County, Arizona.* Flagstaff, Ariz.: Museum of Northern Arizona, 1974.

Bauval, Robert, and Adrian Gilbert. *The Orion Mystery: Unlocking the Secrets of the Pyramids.* New York: Crown Publishers, Inc., 1994.

Bradfield, Richard Maitland. *An Interpretation of Hopi Culture.* Derby, England: Self-published, 1995.

Cunkle, James R. *Raven Site Ruin: Interpretive Guide.* St. Johns, Ariz.: White Mountain Archaeological Center, n.d.

Gilbert, Adrian. *Signs in the Sky.* London: Bantam Press, 2000.

Hancock, Graham, and Robert Bauval. *The Message of the Sphinx: A Quest for the Hidden Legacy of Mankind.* New York: Three Rivers Press, 1996.

Hancock, Graham, and Santha Faiia. *Heaven's Mirror: Quest for the Lost Civilization.* New York: Crown Publishers, Inc., 1998.

Haney, Mark A. *Skyglobe 2.04 for Windows.* Ann Arbor, Mich.: KlassM Software, 1997.

Houk, Rose. *Sinagua: Prehistoric Cultures of the Southwest.* Tucson, Ariz.: Southwest Parks and Monuments Association, 1992.

James, T. G. H. *Ancient Egypt: The Land and Its Legacy.* Austin, Tex.: University of Texas Press, 1989.

Johnson, Ginger. *A View of Prehistory in the Prescott Region.* Prescott, Ariz.: Self-published, 1995.

Krupp, E. C. *Skywatchers, Shamans & Kings: Astronomy and the Archaeology of Power.* New York: John Wiley & Sons, Inc., 1996.

Lister, Robert H., and Florence C. *Those Who Came Before:*

Southwestern Archeology in the National Park System. Tucson, Ariz.: Southwestern Parks and Monuments Association, 1994.

Loftin, John D. *Religion and Hopi Life in the Twentieth Century.* Bloomington, Ind.: Indiana University Press, 1994.

Malville, J. McKim, and Claudia Putnam. *Prehistoric Astronomy in the Southwest.* Boulder, Colo.: Johnson Books, 1989.

Noble, David Grant. *Ancient Ruins of the Southwest: An Archaeological Guide.* Flagstaff, Ariz.: Northland Publishing, 1989.

Oppelt, Norman T. *Guide to Prehistoric Ruins of the Southwest.* Boulder, Colo.: Pruett Publishing Company, 1989.

Reid, Jefferson, and Stephanie Whittlesey. *The Archaeology of Ancient Arizona.* Tucson, Ariz.: University of Arizona Press, 1997.

Smith, Stan. "House of the Badlands." *Arizona Highways,* August 1993.

Talayesva, Don, and Leo W. Simmons, eds. *Sun Chief: An Autobiography of a Hopi Indian.* New Haven, Conn.: Yale University Press, 1974.

Waters, Frank, and Oswald White Bear Fredericks. *Book of the Hopi.* New York: Penguin, 1987.

Williamson, Ray A. *Living the Sky: The Cosmos of the American Indian.* Norman, Okla.: University of Oklahoma Press, 1989.

www.teamatlantis.com.

Chapter 13. On the Possibility of Instantaneous Shifts of the Poles

Deimel, R. F. *Mechanics of the Gyroscope: The Dynamics of Rotation.* New York: Dover, 1950.

Gehrels, Tom. "Collision with Comets and Asteroids." *Scientific American,* March 1996.

Hapgood, Charles. *The Path of the Pole*. Philadelphia, Pa.: Chilton Book Co, 1970.

Spedicato, E. "Apollo objects, Atlantis and the deluge: a catastrophical scenario for the end of the last glaciation" Quad. 90/22, 1990, Bergamo University, Italy.

Wetherill, G. W. "The Apollo Objects." *Scientific American,* May 1979.

Chapter 14. Mysterious Strangers

British Museum's commentary, in Rawson, Jessica. *Mysteries of Ancient China*. New York: George Braziller, 1996.

Hancock, Graham. *Fingerprints of the Gods: The Evidence of Earth's Lost Civilization*. New York: Three Rivers Press, 1995.

———. *Heaven's Mirror*. New York: Three Rivers Press, 1998.

Contributors

S. Badrinaryan is the chief geologist with the scientific team from the National Institute of Ocean Technology (NIOT), the Indian Agency responsible for the underwater surveys in the Gulf of Khambhat. His current field of research, in conjunction with archaeologists and other scientists, is exploring the mysterious Harappan (Indus Valley) civilization that flourished across what is now Pakistan and northwest India from about 3000 BCE.

Flavio Barbiero, a retired admiral in the Italian Navy, spent most of his professional life as an engineer in research centers of the Italian Navy and the North Atlantic Treaty Organization (NATO). He has worked in conjunction with various scientific institutions on several research programs ranging from paleoclimatology—with expeditions in Antarctica—to biblical archaeology in Israel. He is the author of several books, including *Una Civiltà sotto ghiaccio (A frozen civilization)*, *Atlantis in Antarctica*, *Bible Without Secrets*, and *The Secret Society of Moses*. He lives in Italy.

Robert Bauval has researched and written extensively about the pyramids of Giza. His Orion Correlation Theory, in particular, has attracted enormous popular attention and support around the world, but has also stirred up a hornet's nest of animosity amongst Egyptologists and orthodox historians. Bauval's latest book, *The Egypt Code,* takes his groundbreaking work even further and provides us with stunning new insights into the mysteries of ancient Egyptian religion and civilization. For more information on Robert Bauval, please visit his website: www. robertbauval.co.uk.

Mark Booth, a London publisher who has taught philosophy and theology at Oxford, is the author of *The Secret History of the World,* which reveals extraordinary and thought provoking insights into the esoteric teachings of secret societies down through the ages, and offers a radically new (or perhaps very ancient) perspective on human history. *The Secret History of the World* was published by Quercus Books, London, in 2007.

Andrew Collins is a science and history writer and the author of various books that question the way we perceive the past. These include *From the Ashes of Angels* (1996), *Gods of Eden* (1998), *Gateway to Atlantis* (2000), and *The Cygnus Mystery* (2006). He lives near Marlborough, Wiltshire, in the United Kingdom. For more information about Andrew Collins and his work please visit www .andrewcollins.com.

Scott Creighton is fascinated by the mysteries of the ancient world. Out of this fascination he has developed a series of extraordinary theories that challenge entrenched orthodox ideas of history, particularly those concerning the design and purpose of the pyramids at Giza. An ICT engineer by profession, he lives and works in Scotland.

Walter Cruttenden is the Director of the Binary Research Institute, an archaeo-astronomy think tank located in Newport

Beach, California. His focus is on the astronomy, mythology, and artifacts of ancient cultures. Cruttenden is the author of *Lost Star of Myth and Time,* published by St. Lynn's Press, and the writer and producer of the award-winning documentary *The Great Year,* which examines the myth and folklore of ancient cultures and seeks to find the message that these cultures left for modern man. He resides in California.

Gary A. David is an independent researcher and the author of *The Orion Zone: Ancient Star Cities of the American Southwest.* This work reveals how the constellation Orion provided the template by which the Anasazi (ancestral Hopi) navigated their migrations across the southwestern United States, migrations that took place over thousands of years.

Graham Hancock is the author of the major bestsellers *The Sign and the Seal, Fingerprints of the Gods, Keeper of Genesis, Heaven's Mirror,* and *Supernatural.* His books have sold more than five million copies worldwide and have been translated into twenty-seven languages. His public lectures and broadcasts have further established his reputation as an unconventional thinker who raises controversial questions about humanity's past.

Richard C. Hoagland has done more than any other researcher to draw attention to the devastating possibility that the ruins of ancient and unknown civilizations may lie on the surface of the planet Mars and on the moon. Although his views and his meticulous presentation of the evidence have been pilloried by many mainstream scientists, he has continued for many years to build up his case and put it before the public, drawing attention to what he believes are deeply disturbing cover-ups and misrepresentations of the facts by NASA (the U.S. National Aeronautics and Space Administration).

Glenn Kreisberg is a radio frequency engineer, writer, and researcher who currently serves as vice president of the New England Antiquities Research Association (www.NEARA .org). He has researched and published articles and interviews on subjects as diverse as electromagnetism and the ancients, historic bluestone quarrying in upstate New York, and lithic sites and alignments in northeast America. He is also the founder and editor of the alternative science and history website www .ASHnews.org., and the current editor of the Author of the Month page featured at www.grahamhancock.com.

Edward F. Malkowski has a lifelong interest in ancient history, with a special interest in philosophy, the rise of civilization, and the development of ancient religious beliefs. In the late 1990s he began investigating the possibility that the ancient biblical stories in Genesis were based on historical fact, which led to his first book *Sons of God—Daughters of Men.* Other books followed; they include *Before the Pharaohs, The Spiritual Technology of Ancient Egypt,* and his most recent work, *Ancient Egypt 39,000 BCE.*

Robert M. Schoch, who earned his Ph.D. in geology and geophysics at Yale University, has been a full-time faculty member of the College of General Studies at Boston University since 1984. He became world famous in the early 1990s for his work on the Great Sphinx, which demonstrated that the statue may be much older than the dynastic civilization of ancient Egypt. Schoch has authored, coauthored, and edited many books, including *Voices of the Rocks, Voyages of the Pyramid Builders, Pyramid Quest,* and *The Parapsychology Revolution.*

John Anthony West is a writer, scholar, and Pythagorean, born in New York City. He is the author of *The Traveler's Key to Ancient*

Egypt and *Serpent in the Sky: The High Wisdom of Ancient Egypt,* an exhaustive study of the revolutionary Egyptological work of the French mathematician and Orientalist R. A. Schwaller de Lubicz. John Anthony West won an EMMY Award for his 1993 NBC Special Documentary *The Mystery of the Sphinx,* hosted by Charlton Heston. He is featured in the recent 8-Episode DVD series *Magical Egypt: A Symbolist Tour.*

BOOKS OF RELATED INTEREST

Forbidden History
Prehistoric Technologies, Extraterrestrial Intervention,
and the Suppressed Origins of Civilization
Edited by J. Douglas Kenyon

Forbidden Science
From Ancient Technologies to Free Energy
Edited by J. Douglas Kenyon

Black Genesis
The Prehistoric Origins of Ancient Egypt
by Robert Bauval and Thomas Brophy, Ph.D.

The Suppressed History of America
The Murder of Meriwether Lewis and the Mysterious
Discoveries of the Lewis and Clark Expedition
by Paul Schrag and Xaviant Haze

Atlantis beneath the Ice
The Fate of the Lost Continent
by Rand Flem-Ath and Rose Flem-Ath

The Giza Prophecy
The Orion Code and the Secret Teachings of the Pyramids
by Scott Creighton and Gary Osborn

Lost Technologies of Ancient Egypt
Advanced Engineering in the Temples of the Pharaohs
by Christopher Dunn

Advanced Civilizations of Prehistoric America
The Lost Kingdoms of the Adena, Hopewell,
Mississippians, and Anasazi
by Frank Joseph

Inner Traditions • Bear & Company
P.O. Box 388
Rochester, VT 05767
1-800-246-8648
www.InnerTraditions.com

Or contact your local bookseller